Deserts: A Very Short Introduction

VERY SHORT INTRODUCTIONS are for anyone wanting a stimulating and accessible way into a new subject. They are written by experts, and have been translated into more than 45 different languages.

The series began in 1995, and now covers a wide variety of topics in every discipline. The VSI library now contains over 500 volumes—a Very Short Introduction to everything from Psychology and Philosophy of Science to American History and Relativity—and continues to grow in every subject area.

Titles in the series include the following:

Nick Middleton

DESERTS

A Very Short Introduction

OXFORD
UNIVERSITY PRESS

OXFORD
UNIVERSITY PRESS

Great Clarendon Street, Oxford, OX2 6DP,
United Kingdom

Oxford University Press is a department of the University of Oxford.
It furthers the University's objective of excellence in research, scholarship,
and education by publishing worldwide. Oxford is a registered

© Nick Middleton 2009

The moral rights of the author have been asserted
Database right Oxford University Press (maker)

First published 2009

All rights reserved. No part of this publication may be reproduced, stored in
a retrieval system, or transmitted, in any form or by any means, without the
prior permission in writing of Oxford University Press, or as expressly permitted
by law, by licence or under terms agreed with the appropriate reprographics
rights organization. Enquiries concerning reproduction outside the scope of the
above should be sent to the Rights Department, Oxford University Press, at the
address above

You must not circulate this work in any other form
and you must impose this same condition on any acquirer

Published in the United States of America by Oxford University Press
198 Madison Avenue, New York, NY 10016, United States of America

British Library Cataloguing in Publication Data
Data available

Library of Congress Cataloging in Publication Data
Data available

ISBN 978-0-19-956430-9

Printed and bound by
CPI Group (UK) Ltd, Croydon, CR0 4YY

This book is for Mia

Contents

List of illustrations

List of illustrations

Introduction

Deserts embrace a huge variety of conditions. There are hot deserts and cold deserts, sandy deserts and rocky deserts, foggy deserts and sunny deserts, barren deserts and lush, verdant parts of deserts. Deserts are found at high altitudes and below sea level, on coastlines and deep inside continental interiors. They are places of romance and of hardship, of deep spirituality and dazzling natural beauty. Deserts include vast areas of stark wilderness and large, densely populated cities with every modern amenity.

People have been inextricably linked with deserts since time immemorial. Crucial early developments in agriculture and the rise of the city took place in the deserts of the Middle East, the 'cradle of civilization'. The world's deserts have been perceived as places of magnificence and mystery, silence and contemplation, inspiring novelists, poets, artists, and film-makers, and acting as the birth place for three of the world's great religions.

These are places symbolized by Nature's extremes. Deserts are the driest and hottest environments on Earth. They contain vast fields of sand and huge, desiccated lakes of salt; cacti that can live for a thousand years; and frogs that spend most of their lives in a state of suspended animation. For much of the time a desert landscape seems desolate and lifeless, but a rainstorm can turn the same territory into a vibrant tableau bursting with life, a pulse of abundance reflecting the position of water as the desert's primary limiting resource.

The irregular dynamics of desert life can make these environments seem marginal in many respects. At a time when the implications of global climate change weigh heavily on many people's minds, deserts should be places of particular interest. Their character means that even small shifts in climatic factors can have numerous direct and indirect effects on all aspects of a desert landscape and the people who make a living there. Deserts are among the regions most vulnerable to climate change, but their great global extent also means that an understanding of how deserts respond to change has implications for the planet as a whole and thus for every one of us.

Everybody has a notion of what a desert should look like. Effective definitions of deserts vary according to the background of those doing the defining and the purpose of their enquiry. An artist's approach to deserts may be different from the stance taken by a scientist although, broadly, the two usually overlap geographically. It may, or may not, be surprising to learn that no universally accepted definition of the term 'desert' exists.

Many believe it is one of the oldest written words, originating in an ancient Egyptian hieroglyph pronounced 'tésert' and filtered through the Latin verb *deserere* – to abandon or forsake – to reach its meaning today. This is made up of notions of lifelessness, a scarcity of plant and animal life, combined with and stemming from arid conditions, or a general lack of water.

Numerous scientists have tried to define desert regions precisely and draw their boundaries on maps. These attempts all identify the same core desert areas, but sometimes disagree about where desert boundaries lie. To some extent, this is a problem inherent in any attempt to draw fixed boundaries on maps of natural systems. We are able to designate sharp frontiers that separate countries and other political areas because these boundaries were created by people, but most of Nature's edges actually grade imperceptibly from one zone to the next. Many are also constantly in a state of flux.

Aridity is typical of deserts, and probably the most common condition used to classify them. These classifications employ climatic data, the most straightforward of which is rainfall, but rainfall is only half the story. A much better measure of aridity includes rainfall and other forms of moisture input – including dew, fog, and snow – but also the loss of water by evaporation and by transpiration from plants.

In its simplest form, this approach to the 'water balance' of an area can identify an arid region as 'empty bucket land': if you leave a bucket on the ground and it never fills up, you are in a desert. In more technical language, an arid zone is where the supply of water by all forms of precipitation is exceeded by the water lost via evaporation and transpiration.

This ratio of water inputs to outputs has been used to create a number of forms of aridity index. The most complete forms also include an assessment of changes in the amount of water stored in soils, beneath the ground surface. This approach can be used to identify and quantify degrees of aridity. Hence, 'hyper-arid', 'arid', and 'semi-arid' are terms commonly used in discussions of deserts. 'Dry sub-humid', another climatic subdivision, is a further category sometimes used. Each can be strictly determined using an aridity index. However, the divisions between these categories vary, depending on the scheme used.

To an extent, all such divisions are arbitrary, so some prefer to talk about 'arid zones' or 'drylands'. Both of these terms are more or less synonymous with the words 'desert' and 'semi-desert' combined. In this book, the term 'desert' is used for the most part to denote those areas with a hyper-arid or arid climate. Areas with climates categorized as semi-arid or dry sub-humid are referred to here as semi-desert or desert fringe. Unfortunately, like all of the attempts to categorize deserts, this apparently straightforward rule of thumb is not perfect. Much of southern Africa's Kalahari desert, for instance, has a semi-arid climate.

Deserts can be further subdivided climatically into those with cold winters (such as the Central Asian deserts, the Patagonian desert in South America, and some of the North American deserts); coastal deserts where fog is common (for example, the Atacama and the Namib deserts); and hot deserts where winter frosts are unusual (such as the Sahara, Arabia, and the Australian deserts).

Climate is by no means the only way of identifying and demarcating deserts. Vegetation and soils are two others, though both are related to climate. One significant problem with using climatic information to delimit deserts is the relative lack of climatic data for desert regions. The general deficiency of permanent settlements in deserts means that they host relatively few weather stations. We have much better coverage of the Earth's surface with good satellite imagery of vegetation. Most people agree that deserts are large areas with little vegetation cover and a great expanse of bare soil. In biological terms, deserts are places that contain few plants and animals, but those that are found have clear adaptations for survival in arid conditions.

None of these approaches to desert definition is foolproof. All have their advantages and drawbacks. However, each approach delivers a global map of deserts and semi-deserts that is broadly similar to Figure 1, and this map shows the subject areas of this book. They make up a large proportion of the world's land surface; nearly half in fact. Roughly, deserts cover about one-quarter of our planet's land area, and semi-deserts another quarter.

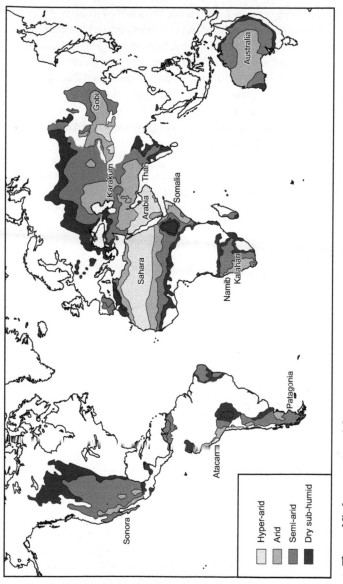

1. The world's deserts and semi-deserts

The world's deserts are found in two main east–west belts that run parallel to the Equator, in the northern and southern hemispheres. North of the Equator, these deserts are the Mojave, Sonoran, Great Basin, and Chihuahuan deserts in North America; the Sahara in North Africa and the Somali-Ethiopian deserts in the Horn of Africa; the deserts of Asia, including the Arabian, Iranian, and Thar deserts stretching from the Middle East into Pakistan and India, as well as the deserts of Central Asia, and the Taklimakan and Gobi deserts in China and Mongolia.

In the southern hemisphere, the desert belt is formed by South America's Atacama, Altiplano-Puna, Monte, and Patagonian deserts; the Namib, the Kalahari, and the Karoo in southern Africa; and the large expanse of the Australian deserts.

Immediately, some further explanation is required. Figure 1 demonstrates that this book will not cover other biological deserts, such as large expanses of ice in polar and glacial regions and subtropical ocean gyres, though both are notable for their low biological activity, relative lack of nutrients, and low stocks of organisms.

Polar and other ice-covered regions could easily be called deserts, but their environments are sufficiently distinctive for most people, including myself, to put them into a separate category. For the same reason, the least biologically productive regions of the oceans, the subtropical gyres, are excluded.

This book is not about marine deserts, although, interestingly, large regions of the world's oceans are considered to have an arid climate. Details are scant, because of the lack of good long-term meteorological information for oceans, but considerable areas in the subtropics, most to the west of the continents, probably receive precipitation totals of 200 millimetres a year or less due to the prevailing atmospheric stability in these areas. The climates of desert islands, such as Cape Verde in the North Atlantic and

Socotra in the Arabian Sea, are affected in the same way. These islands with desert climates, and others like them, are within the scope of this volume.

Some deserts have already been mentioned by name, but different authorities sometimes use names in different ways. The Sahara is the world's largest desert, covering some 9 million square kilometres and extending across the territory of 10 countries in North Africa. However, parts of the Sahara are often given other labels, such as the Ténéré desert, the Libyan desert, the Egyptian desert, the Nubian desert, and the Bayuda desert.

The Atacama is the world's longest expanse of desert, extending along more than 2,500 kilometres of South America's Pacific coastline in Peru and Chile. Some call the Peruvian section the Sechura desert. Within Peru, the name Sechura tends to be confined to the northernmost section of dry Peruvian coast. Others call the whole lot the Peruvian-Chilean desert.

A more complex set of local names occurs in the Thar desert, which spans the border between Pakistan and India. In Pakistan, the southern part of the Thar is known as the Sind desert, or the Parkar Thar, and the northern part is the Cholistan desert. However, the Cholistan is also locally known as Rohi and sometimes called the Nara desert. Another area, to the northwest of the Thar in Baluchistan, is called the Thal desert. In India, the Thar is included within a larger dryland area variously known as the Rajasthan or Rajputana desert, or the Great Indian desert.

Australia has a larger proportion of its area occupied by deserts than any other continent: about 70% is classified as arid or semi-arid. This area has numerous local names. These deserts include the Great Victoria, the Great Sandy, the Tanami, the Simpson, the Gibson, the Little Sandy, and the Strzelecki, each of which occupy at least 1% of the national land area. Australia's smaller deserts include the Sturt Stony, the Tirari, and the Pedirka.

2. Sana'a, the desert capital of Yemen, which has been inhabited for more than 2,500 years

These brief sorties into the names used in a few locations serve to illustrate the great diversity of deserts worldwide. This book initially approaches the subject from a physical science perspective, hence it is appropriate to sum up how deserts are identified. They are regions of great aridity, large areas of bare soil, very little vegetation cover, and few animals. The wildlife that is found in deserts has obvious adaptations to lengthy periods when minimal water is available.

This book aims to throw some light on these and other inherent qualities of desert regions. All too often, they are areas depicted as sterile, drought-prone wastelands, but this largely negative view is both unfortunate and imbalanced. The world's deserts also boast a rich biodiversity, a long history of successful human occupation, and a great importance for the workings of our planet in general. They harbour immense natural beauty, extraordinary adaptations, diverse cultures, and great civilizations.

Chapter 1
Desert climates

Deserts have their aridity in common, but this shared feature should not obscure the fact that a wide variety of conditions is encompassed by the term. Deserts are not homogeneous and not all deserts are the same.

The climate of a particular region is made up of the average weather conditions over a period of a few decades. It is defined by variations in factors such as temperature, rainfall, wind, and sunshine. High temperatures and a paucity of rainfall are two aspects of climate that many people routinely associate with deserts, and indeed both the world's hottest and driest places are located in desert areas.

However, desert climates also embrace other extremes. Many arid zones experience freezing temperatures and snowfall is commonplace, particularly in those situated outside the tropics. Although dry for most of the year, individual storms can bring large amounts of rainwater to deserts. For much of the time, desert skies are cloud-free, meaning deserts receive larger amounts of sunshine than any other natural environment. However, fog is not unusual in several coastal deserts, and some typically have more foggy days than rainy days.

The combinations of climatic factors and their variability are almost limitless. This chapter might have been entitled 'the desert climate' but the plural form is deliberate: there is no one desert climate.

Causes of aridity

Deserts are located where they are because some places have naturally low rainfall or have temperatures high enough to evaporate what water there is. Four main planetary-scale factors dictate these characteristics and hence the global position of deserts.

Atmospheric stability

Large areas of gently descending, stable air are responsible for the two bands of deserts centred roughly 30° north and 30° south of the Equator. These are the subtropical high pressure belts which explain the location of the Sahara, the Arabian deserts, the Kalahari, and the Australian deserts.

Continentality

Distance from the oceans prevents rain-bearing winds from penetrating the interior of large continents. Most of the water vapour in the world's atmosphere is supplied by evaporation from the oceans, so the more remote a location is from this source the more likely it is that any moisture in the air will have been lost by precipitation before it reaches continental interiors. The deserts of Central Asia illustrate this principle well: most of the moisture in the air is lost before it reaches the heart of the continent which includes the Karakum desert, the Kyzylkum, and the Taklimakan.

Rain shadows

Aridity is intensified in deserts sheltered from prevailing moist winds by a mountain range. The air is forced up the mountain, it cools, and its moisture forms clouds and falls as rain. On the far side of the mountain, the winds are thus much drier and the downwind area is said to be in a 'rain shadow'. In Australia, the Great Divide creates a rain-shadow effect for the continent's central deserts, and the Sierra Nevada creates the same effect for some of the North American deserts.

Cold ocean currents

Cold currents at the surface of oceans have a desiccating effect on coastal margins of adjacent continents because cold seas produce relatively little atmospheric moisture by evaporation, so limiting the moisture in the atmosphere which can fall as rain. Good examples of these currents, cold because they flow from the poles towards the tropics or because they are formed by water welling up from great depth, include the Benguela Current along the Atlantic coast of southern Africa (affecting the Namib desert), and the Peru or Humboldt Current along the Pacific coast of South America (the Atacama desert).

3. This view looking eastward across the high Atacama shows how the Andes create a rain-shadow effect, blocking the flow of moist air from the Amazon

All the world's deserts are affected by at least one of these main causes of aridity on the global scale, and some by several. The reasons for extreme aridity in the Atacama desert, for instance, are related to three key features: (1) its position beneath the warm, dry, descending airflow of the subtropical high pressure belt; (2) the adjacent, northward-flowing, cold oceanic Humboldt Current, which suppresses evaporation and coastal precipitation; and (3) the mountain barrier of the Andes, which creates a rain-shadow effect by blocking any flow of moisture from the Amazon basin to the east.

The hottest place on Earth

Some deserts can be very hot. During the daytime, temperatures can rise rapidly due to clear skies and high levels of sunshine. Few clouds mean that many deserts receive close to the maximum possible amount of radiation from the Sun. In Egypt, for example, the average number of sunshine hours is around 3,500 a year. In Tibet, the annual average is about 3,400 hours.

Consequently, desert interiors can experience air temperatures in excess of 40°C for many consecutive days. These conditions can be unpleasant for people and present a challenge to desert plants and animals. The title of 'hottest place on Earth' has been claimed by several locations, all of them in deserts. Death Valley, in California, USA, held the record for the highest recorded air temperature of 56.7°C from 1913 to 1922, but lost the world record in September 1922 when an air temperature of 58.0°C was recorded at El Azizia in northern Libya. Another place that often appears in compilations of meteorological records is Dallol in the Danakil desert of Ethiopia. Dallol holds the record for the highest average annual air temperature, of 34.5°C.

These records were measured at weather stations using standardized equipment with the temperature taken 1.5 metres above the ground surface. The thermometer is kept inside a wooden box – a Stevenson screen – to protect it from direct solar radiation and from reflected terrestrial radiation. The objective is to measure the temperature of the air circulating in the area, so the Stevenson screen has louvered sides to allow air to pass freely over the instrument.

It is possible that hotter temperatures have occurred elsewhere in desert locations without instruments to record them. Generally, long-term meteorological data for many deserts are deficient simply because of the sparse populations that live in desert areas. Sensors carried on satellites have the advantage over fixed weather stations of being able to collect data with continuous geographical coverage. Attempts to locate the 'hottest place on Earth' using data from satellites have focused on measuring temperature at the Earth's surface, which is typically hotter than the temperature of the air.

Global satellite-derived maps of the highest annual maximum land surface temperatures confirm the fact that this planet's hottest places are in arid, sparsely vegetated landscapes. Large parts of the Sahara, the Middle East, the Gobi, most of Australia, and large areas in western North America stand out as places where land surface temperatures regularly exceed 60°C. The highest land surface temperature documented by satellite was in the Lut desert in Iran, where 70.7°C was recorded in 2005. None the less, direct monitoring of ground surfaces has returned still greater temperatures. The temperature of a bare sand surface at Repetek, a desert research station in the Karakum in Turkmenistan, has been measured at 79.4°C. In the Red Sea hills, north of Port Sudan, a sand temperature of 83.5°C has been recorded.

Temperature range

A clear distinction can be made between deserts in continental interiors and those on their coastal margins when it comes to the range of temperatures experienced. Oceans tend to exert a moderating influence on temperature, reducing extremes, so the greatest ranges of temperature are found far from the sea while coastal deserts experience a much more limited range.

Generally clear, cloud-free skies with a relative lack of moisture allow daytime temperatures in deserts to soar but the same conditions also mean that the contrast between temperatures during the day and at night – the so-called 'diurnal' temperature range – is greater in some deserts than in any other environment. This is particularly so in continental interiors, where diurnal temperature ranges of 20°C are commonplace. In extreme cases, the diurnal range can be much greater and it may vary seasonally. In the Negev desert, the typical diurnal range in July is 23–36°C, and about 6–18°C in January. In Death Valley, USA, a maximum range between day- and night-time temperature of 41°C was recorded in August 1891, and the mean diurnal range for that month was 35°C. The main exception to such large desert diurnal temperature ranges occurs in coastal zones where the close proximity of relatively cool ocean surfaces moderates any extremes. Diurnal temperature ranges of 10°C are more typical in these desert areas.

A contrast in temperatures between summer and winter can be seen in most deserts. Al-Quwaiayh, in central Saudi Arabia, has a subtropical desert climate with relatively cool winters (the average January temperature is 15°C) but summer temperatures that frequently exceed 45°C.

Deserts at relatively high latitudes (such as the Central Asian deserts) and relatively high altitudes (as in Tibet) experience greater seasonal contrasts. The most important determining factor is the seasonal variation in solar radiation at higher latitudes, but the continentality of climate in the Gobi desert, for example, also contributes to seasonal temperature extremes. The annual range in monthly average temperature is greater than 40°C at some meteorological stations in Mongolia and northern China. In summer, temperatures can reach 40°C, and in winter daily average temperatures are commonly −10°C. Extremes are greater still, of course. The difference between the highest and lowest temperature recorded in a single year at Saixan Ovoo in the Mongolian Gobi was 77.7°C (a summer maximum of 36.8°C, winter minimum of −40.9°C) in 2002. Seasonal temperatures can also be extreme in the high-latitude Patagonian desert in Argentina, where winter temperatures in the south are as low as −29°C while summer temperatures can be well over 37°C. Temperature ranges in Tibet are similarly large, both diurnally and seasonally, due both to Tibet's location outside the tropics and to its altitude. The Tibetan Plateau averages out at around 4.5 kilometres above sea level. On the northern part of the plateau, a desert area known as the Changtang, midday temperatures in winter of 1–5°C frequently follow night-time lows of −35°C. The temperature can drop below freezing point (0°C) at night even in the summer, when afternoon highs can reach 40°C.

Frost and snow

Freezing temperatures occur particularly in the mid-latitude deserts, but by no means exclusively so. Frosts are certainly unusual in the hot deserts, but records of minimum temperatures for the Sahara show that large areas have experienced temperatures below 0°C at some point in the last 100 years. Indeed, snowfall occurs at the Algerian oasis towns of Ouagla and Ghardaia, in the northern Sahara, as often as once every 10 years on average. Freezing temperatures have also occurred in many parts of the Arabian Peninsula in spite of the low latitudes.

Frost, and the possibility of snow, increase outside the tropics. In the Sonoran desert in North America, at about 30°N the number of days when temperatures fall below 0°C is around 100 a year on average. In the Patagonian desert, which extends southwards from about 40°S to more than 50°S, the number of frost-free days is generally less than 100 a year, and much of the precipitation occurs as snow, sometimes heavy, during the winter months of

June, July, and August. Altitude obviously has an impact too. In the western parts of the Tibetan Plateau, the average temperature is below freezing on more than 200 days a year.

The very low winter temperatures associated with some continental deserts can present severe problems for their inhabitants. In extreme winter conditions, livestock commonly face difficulties in finding pasture due to a combination of freezing winds, extreme cold, intense snowfall, and extended snowdrifts. These harsh conditions, known as *dzhut* in the semi-deserts of Kazakhstan and *dzud* in the Mongolian Gobi, can result in a large number of livestock deaths.

Mongolian herders distinguish between a white *dzud*, caused by deep and prolonged snow cover that makes fodder inaccessible to animals, and a black *dzud* that occurs during cold weather without snow when surface water sources and the ground freeze, restricting access to both food and water. The impact of a *dzud* can be catastrophic. The mass demise of animals during the winter months of October to May is further intensified if preceded by drought or poor grazing conditions in the previous growing season. A series of devastating *dzuds* that struck Mongolia between 1999 and 2003 was responsible for the death of more than 25% of the country's livestock – some 8.5 million animals – resulting in very widespread economic difficulties for many Mongolians.

Low rainfall

Deserts typically receive low levels of rainfall. They also receive other forms of precipitation, including fog, dew, hail, and snow, but the overall total of precipitation received on average each year is characteristically low. In extreme cases, in hyper-arid zones, such mean annual totals are less than 10 millimetres. Quillagua, a village in Chile's Atacama desert, holds the title of 'driest place on Earth', with an annual average precipitation of just 0.5 millimetres over the period 1964 to 2000. The Atacama is a very dry desert. One hundred years of meteorological data from the town of Iquique gives an annual average precipitation of around 2 millimetres for the 20th century.

Another characteristic of rainfall in deserts is its variability from year to year which in many respects makes annual average statistics seem like nonsense. A very arid desert area may go for several years with no rain at all (Iquique received no precipitation whatsoever in the 1960s). It may then receive a whole 'average' year's rainfall in just one storm (see the table). For example, a period of heavy rainfall over a week in April 2006 at the coastal town of Luderitz in the Namib desert brought a total of 102 millimetres, or about six times its annual average rainfall of 16.7 millimetres. Luderitz received well above its annual average rainfall on each of three days that week. An even greater extreme occurred in the lower Chicama valley in Peru where the average precipitation is only 4 millimetres a year. One storm in March 1925 brought no less than 394 millimetres of rain.

Table: Extremes of precipitation in deserts

Station	Mean annual precipitation (mm)	Highest precipitation in a single storm (mm)	Storm date
Chicama (Peru)	4.0	394.0	March 1925
Luderitz (Namibia)	16.7	102.0	16–22 April 2006
Masirah Island (Oman)	70.0	430.6	13 June 1977
Diego Aracena (Chile)	1.3	9.1	5 July 1992
El Djem (Tunisia)	275.0	319.0	25–27 September 1969
Ziyaratgah (Iran)	28.0	30.0	6 March 1997
Oodnadatta (Australia)	173.4	200.0	9 February 1976
El Cajoncito (Mexico)	220.7	237.5	29 September 1982

Such events can cause severe difficulties for inhabitants of desert regions, particularly due to flash flooding. The heavy rain and strong winds associated with the tropical cyclone that brought a huge 431 millimetres of rainfall in one day to Masirah Island in the Arabian Sea in June 1977 damaged virtually all of the dwellings on the island and left 2,000 families homeless. It would be misleading to suggest that all desert rainfall occurs in such intense storms, but deserts do have relatively few rainy days. Oman averages about 12 rainy days a year. For comparison, it rains on about one day in three in England and no English station has ever received its entire annual average rainfall in a 24-hour period.

Rainfall in deserts is also typically very variable in space as well as time. Hence, desert rainfall is frequently described as being 'spotty'. This spottiness occurs because desert storms are often convective, raining in a relatively small area, perhaps just a few kilometres across. An assessment of the spottiness of rainfall in an extremely arid part of Israel, the southern Arava, part of the Jordan rift valley, concluded that localized, high-intensity storms provided between a half and two-thirds of the total rainfall in the region. The proportion of the region receiving rainfall from one of these storms, which typically last just a few minutes, may be as low as 20%.

Another study of precipitation variability, in the USA's Mojave desert, found wide variations in totals between weather stations. The desert area studied is just 150,000 square kilometres, or about 375 × 375 kilometres, and has an overall annual average precipitation of 137 millimetres. The highest annual average of the 52 stations was 310 millimetres and its lowest 34 millimetres.

Normally, precipitation in deserts is far exceeded by evaporation, the basic premise of arid lands. Of course, evaporation only occurs when there is water to evaporate and since this is rarely the case in deserts, climatologists have developed the idea of 'potential evaporation'. This is the amount of evaporation that would occur given a never-ending exposed area of water to be evaporated.

Deserts have some of the highest potential evaporation rates of any environment. Al-Ain in the United Arab Emirates receives on average 78 millimetres in annual precipitation with an annual potential evaporation of 2,000 millimetres. In the Badain Jaran desert of northern China, where the mean annual precipitation is less than 90 millimetres, annual potential evaporation is over 2,500 millimetres. Some of the highest mean annual rates are found in the world's driest deserts. The highest potential evaporation rates in the Atacama exceed 3,500 millimetres a year. In the southern part of the Libyan desert in south-central Egypt, where rainfall averages 0 to 5 millimetres annually, the annual potential evaporation may reach 5,000 millimetres.

Dew and fog

In some deserts, rainfall is not the most important type of atmospheric precipitation. Dew and fog are two forms of moisture that occur during the early morning and are evaporated by daytime heating. Both phenomena often take place more frequently and regularly than rainfall and in some cases deliver more water.

The atmosphere in even the driest deserts contains some moisture as water vapour. This atmospheric moisture is condensed out to appear as droplets of water when the air temperature is lowered to a point at which the air can no longer hold all of its water vapour. Dew is the condensation of atmospheric moisture on an object, such as a leaf or a rock. Fog is the condensation of atmospheric moisture in the air near the ground surface; it is essentially a low-level cloud.

The occurrence of dew and fog contrasts with the sporadic and often unpredictable nature of desert rainfall because precipitation of dew and fog can be much more frequent and reliable, even if the amounts of water precipitated tend to be low. On the Avdat farm, in the heart of the Negev desert in Israel, dew or fog occurs on 195 mornings a year on average, supplying a mean annual yield of 33 millimetres of water.

Similarly high frequencies of fog occur in the deserts on the west coasts of southern Africa and South America. In southwestern Africa, fog is formed in air cooled as it flows over the ocean surface, which is relatively cold thanks to the upwelling Benguela Current. This coastal sea fog is known as *cacimbo* in Angola, *nieselregen* in Namibia, and *Mal-mokkie* in Namaqualand.

The area around Swakopmund is the foggiest part of the Namib desert and has more than 100 fog days per year. The fog reaches more than 100 kilometres inland. Rainfall totals in the central Namib average around 10–20 millimetres a year, but water precipitated from fog is in some cases greater. At Swakopmund, on the coast, the average annual precipitation by fog is 34 millimetres, nearly two times the average annual rainfall of 18 millimetres. Records of moisture received from three different atmospheric processes over one year at Gobabeb, about 55 kilometres from the sea in the Namib, show fog deposition occurring on 67 nights, dew on 53 nights, and rain falling on 12 days. This gives a total of 132 days on which the soil surface was briefly wet. The amount of water produced by fog or dew on a particular night tends to be smaller than that delivered by the average rainy day, but in some years the amount of precipitation deposited by fog at Gobabeb is greater than the amount produced by rain.

Fog that occurs along the coastline of northern Chile is known as *camanchaca*, while on the desert coast of Peru they call it *garua*. Some of these fogs are formed in the same way as in the Namib, but other types of fog also occur. In some cases, fog results from clouds generated over the Pacific Ocean far from the coastline, which are simply transported by the wind to the Atacama. At other times, fog is formed as air is cooled on being forced up highland slopes. On some parts of the Atacama coast, fogs can occur on as many as 190 days a year.

There have been few studies of dew in most of the world's deserts, but given the typically high diurnal temperature ranges of many arid environments it is probably safe to say that some dew occurs in all deserts. Fog is less common, but it is a regular feature of the climate in other coastal deserts flanked by relatively cool ocean currents, including Mexico's Baja California, the coast of Mauritania, and the southeast coast of the Arabian Peninsula from Oman through Yemen and into Saudi Arabia along the Red Sea coast. Fog on the Arabian Peninsula coastline is distinctly seasonal, being associated with the Indian Ocean's southwest monsoon between June and September.

Similar conditions do also occur in continental deserts, though with lower frequency. Mist, effectively a light fog, is observed in the Karakum and Kyzylkum deserts on 10 to 20 days a year on average.

Wind and dust

Hot, dry, and dusty winds are another typical feature of desert climates. Although on average desert winds are no stronger than winds that blow in other environments, some specific regional winds are notable. These include the winds of the high-altitude deserts. The plateau of Tibet and the Altiplano-Puna plateau of the Central Andes are swept by fierce, persistent westerlies that blow high above the surface over most of the Earth.

Another of the strongest and most persistent wind systems on Earth are the low-level winds that blow during the summer monsoon in the Indian Ocean. Their effects are well known in the Horn of Africa, where gale-force winds, known as the *Kharif* in the Somali-Chalbi desert, are common throughout the summer months from June to September. The port of Berbera in northern Somalia (Somaliland) experiences more than 50 days with gales during this period.

The extremely strong winds batter the desert archipelago of Socotra, 100–200 kilometres east of Somalia and some 500 kilometres south of the Arabian Peninsula. The islands were traditionally cut off from the mainland during the summer monsoon because the winds made the passage impossible for boats, although an airport opened on the main island in 1999.

Strong desert winds often pick up particles of sand and dust from ground surfaces that are typically dry and bare. The resulting sand and dust storms are another typical feature of desert climates, and all deserts have at least one and often several named dusty winds that blow in particular seasons. The fierce, desiccating effects of

4. Saharan dust blowing over the Mediterranean from northern Libya. In this satellite image, the thickest plume is moving east-northeast from near Libya's border with Tunisia. Farther east, a thick pall billows northward and thinner plumes blow toward the northwest

the Irifi, an easterly that blows in the Western (formerly Spanish) Sahara, is illustrated by an occasion in March 1941 when the onset of the wind raised the temperature from 18.3°C at noon to 42.8°C by 4pm the same day and maintained a temperature above 39°C for the following 36 hours.

The Harmattan is perhaps the Sahara's best-known dry, dusty wind. It is a feature of the winter months, from November to March, when virtually all parts of West Africa south of the Sahara are under the influence of the northeast trade wind which transports Saharan dust into most parts of the region and beyond, over the Atlantic. For the residents of West Africa, it is a hazardous fact of life. The reduced visibility can cause problems for road and air traffic. Flights are commonly delayed, diverted, or cancelled during severe Harmattans, and accidents sometimes occur. The desiccating conditions also bring a high risk of bush fires. Breathing fine Harmattan dust particles is recognized as a health hazard and can aggravate respiratory infections, pneumonia, and bronchitis. The annual meningitis epidemics in West Africa which affect up to 200,000 people are also closely related to the Harmattan season in their timing.

A similar seasonal airflow raises desert dust over southern Iraq and Kuwait during the summer months from late May to early July. This is the Shamal, which carries the dust in suspension out over the Arabian Gulf as far as the United Arab Emirates. Visibility in Kuwait during a Shamal can fall below 100 metres in extreme cases, forcing the closure of the Emirate's three ports and disrupting air traffic.

More localized dusty winds in many deserts are generated by dry thunderstorms which blow a very discrete wall of dust in front of them across the landscape. These thunderstorms are often called haboobs, the name used in the Sudanese Sahara (from the Arabic *habb*, to blow). Haboobs frequently present a severe danger to motorists and pilots due to the sudden loss of visibility, sometimes to as low as 50 metres, at the dust wall.

Many local warm, dry, and dusty winds blow down mountainsides in desert regions. Examples include the Zonda wind which descends the eastern slopes of the Andes into the Monte desert of northern Argentina, and the Afghanets which originates in the Hindu Kush and blows across the Kyzylkum desert in Uzbekistan.

Microclimates

Most of this chapter on desert climates has been concerned with variations in climatic parameters at a regional scale. Climates can vary over a wide range of spatial scales, however. Changes in temperature, wind, relative humidity, and other elements of climate can be detected over short distances, and this variability on a small scale creates distinctive climates in small areas. These are microclimates, different in some way from the conditions prevailing over the surrounding area as a whole.

At the smallest scale, the shade given by an individual plant can be described as a microclimate. Over larger distances, the surface temperature of the sand in a dune will frequently be significantly different from a nearby dry salt lake because of the different properties of the two types of surface. Over say 100 metres or so, great differences in temperature can occur between opposite north- and south-facing slopes of a hill. In the northern hemisphere, this difference in aspect usually means that the soils on north-facing slopes will remain moist for longer after a rainstorm in comparison to south-facing slopes because of the lower evaporation rates created by fewer hours of direct sunshine.

Microclimates are important because they exert a critical control over all sorts of phenomena. These include areas suitable for plant and animal communities to develop, the ways in which rocks are broken down, and the speed at which these processes occur.

The desert mirage and other tricks of the light

High temperatures and dust in the air, two common characteristics of desert climates, can also combine to produce an interesting range of optical phenomena. Perhaps the best known of these is the desert mirage, an apparent pool of cool, clear water in an otherwise barren environment. It is an illusion much loved by Western film-makers, symbolic of the desert's harsh conditions: a sorry band of French Legionnaires, perhaps, lost in the Sahara desert, parched with thirst and taunted by the illusion of fresh water.

Although the optical illusion is in one sense the product of a fatigued mind, the mirage is, none the less, a real image and can be photographed. What the eye sees and the mind interprets as water are actually rays of light from a blue sky and wisps of cloud that are refracted through air near the ground surface so that they appear to have come from the ground. The illusion occurs anywhere with a steep temperature gradient in the first metre or so above the ground surface. The mirage occurs because the rapid change in temperature also creates rapid differences in the density of the air immediately above the ground, causing refraction of the Sun's rays. A difference of about 4°C between the ground surface and the air 1 metre above will be enough to produce a distinct mirage.

Atmospheric effects also produce spectacular changes in the colour of desert landscapes at different times of day. The colour of Uluru, the mighty sandstone hill in Central Australia also known as Ayers Rock, can change dramatically between midday, when it appears yellow-brown, and at the end of the day, when it appears orangey-red. The contrast is caused by particles of dust and water vapour in the atmosphere that act as a filter, removing bluer light from the incoming rays of the Sun. This filtering effect is less pronounced in the middle of the day when the Sun's rays pass through a small thickness of atmosphere, but at sunrise and

5. Dust and water vapour in the desert air can create some interesting optical effects. Uluru, one of only a few isolated hills in Central Australia, appears to change colour dramatically between noon and sunset

sunset, when the low-angled rays pass through a much greater thickness of atmosphere, the filtering effect is enhanced, making Uluru glow fiery red.

The rhythm of climates

The Earth's weather has never been static. One day may bring dew, the next not. Many deserts experience more or less predictable variations over an annual cycle, most notably in terms of seasonal change in elements such as temperature and precipitation.

Hyper-arid deserts are the main exception, particularly with regard to rainfall. The extremely arid heart of the Sahara, for example, receives so little rainfall that no 'rainy season' can be identified. A similar statement applies to the Atacama. Temperatures in the Atacama are also remarkably uniform throughout the year, with monthly mean air temperatures between 14°C and 16°C. Temperatures in the deserts of the Horn of Africa, which are close to the Equator, also show little variability from month to month.

Less arid deserts, and less arid parts of deserts, do exhibit seasonality. Seasonality becomes more pronounced in deserts outside the tropics. Compared to the central Sahara, the climate of the Kyzylkum desert is very seasonal. Temperatures in the Kyzylkum reach 45°C in July and fall to −25°C in January. Rainy seasons are identifiable in arid and particularly semi-arid desert areas. On the southern margins of the Sahara, the Sahel has a very distinct rainy season which becomes longer and more pronounced with increasing distance from the Sahara. Zouar in northern Chad, in the Sahara proper, receives virtually all its meagre rainfall in July and August. The rainy season at Abeche, on the southern margin of the Sahel in Chad, begins earlier – in May/June – and continues until September.

Precipitation in the deserts of Iran and Afghanistan is similarly very seasonal, occurring almost exclusively in winter and spring (November to May), brought largely by cyclonic disturbances from the Mediterranean. This season is also the time of lowest temperatures in these deserts. In Kerman, Iran, the average daily maximum temperature in January is about 12°C; in July, it is more than 35°C. Further east, in the Indian part of the Thar desert, the mean maximum temperature during the summer is commonly 40°C to 45°C, but in some years it can reach up to 50°C. During the coldest month, January, the normal minimum temperature is between 3°C and 10°C.

The level of temperature prevailing when precipitation occurs is important for an area's water balance and its degree of aridity. A rainy season that occurs during the warm summer months, when evaporation is greatest, makes for a climate that is more arid than if precipitation is distributed more evenly throughout the year. The annual average precipitation in parts of Western

Australia is only about 250 millimetres, but it falls in the cooler season, so enabling wheat to be grown. This situation stands in sharp contrast to the climate of Doleib Hill in central Sudan where most of its 760 millimetres of annual average precipitation falls in the summer months when the average temperature is more than 26°C. In consequence, this area of Sudan has a degree of aridity comparable to parts of North Africa that receive only one-tenth the amount of precipitation.

Variability in desert climates is detectable over longer periods, too. Drought can occur in virtually any climate, but it is perhaps most critical in desert margins, such as the Sahel, where pastures and crops depend on rainfall in one season. Drought can be understood in several different ways. At a simple level, it could be described as less rainfall than expected. At its most fundamental, drought means there is a lack of sufficient water to meet normal requirements in a particular place or at a particular time. It is worth noting, then, that drought to one section of the community (for example, a market gardener) may not be regarded as drought by another (a camel herder). It is also good to realize that the word has no real meaning in hyper-arid deserts, which are almost always dry.

Despite the lack of a completely satisfactory definition, there is no doubt that droughts can result in great hardship for the inhabitants of deserts. Droughts can cause crop failure and the loss of livestock; they can also trigger other natural hazards, such as fires and dust storms. The timing of droughts has always been difficult to predict, but in some deserts climatologists have identified links with the El Niño-Southern Oscillation (shortened to ENSO), a phenomenon of changes in the distribution of warm and cold water in the tropical Pacific Ocean.

These ENSO events have knock-on effects on the weather all over the planet. Droughts in dry areas as far apart as the Sahel, northeastern Brazil, and Australia have all been linked in this way. An ENSO event does not induce drought in all desert areas, however. ENSO effects in the coastal deserts of Peru and Chile are often just the reverse. The relatively warm surface waters bordering these coasts during an ENSO create unusual rainstorms.

Droughts affected by ENSO tend to last not much longer than a year, but other droughts can extend over several years. Many of the worst droughts in Australia, for example, have occurred when one or two very dry years follow several years of generally below-average rainfall. The country's 'Federation drought' began in 1895 and reached its climax in late 1901 and 1902. The history of rainfall in Australia, like other desert areas, also features several periods of a decade or longer that have been distinctly prone to drought. Rainfall in most years is below the long-term average during these periods, which are frequently punctuated by runs of years with recurrent drought. Most of the country had generally low rainfall during the late 1920s and the 1930s, for instance, continuing through most of the 1940s over the eastern states.

Changing climates

When periods of markedly different climate endure for several decades or longer, climatologists start to consider whether they should refer to an actual change in climate rather than a fluctuation. A trend of decreasing rainfall and devastating droughts in the Sahel region during the last three decades of the 20th century has been recognized by climate researchers as one of the largest recent climate changes anywhere in the world. In many desert regions, noticeable trends have also been recorded in temperatures over the last hundred years or so, in line with the general warming of our planet.

Monthly extreme minimum temperatures have increased throughout the Sonoran desert, for example. In Arizona, the number of days when temperatures fell below 0°C was around

115 to 130 in the 1920s, 1930s, and 1940s, but the length of the frost-free season increased towards the end of the century, so that in the 1980s and 1990s temperatures at or below freezing were recorded on 95 to 105 days. Similar warming trends have occurred in many other deserts. In Tibet, the rate at which temperatures have risen has been greater than the global average, and there is also a tendency for the warming trend to increase with altitude, suggesting that the Tibetan Plateau is one of the world's most sensitive areas to respond to global climate change.

Changes in the climate of deserts have also been identified deeper into the past. To make up for the lack of direct weather observations, all sorts of indirect lines of evidence have been used to assess what climates were like at various times in Earth's history. Ancient inscriptions, government and commercial records, diaries, and correspondence can all contain details of particular weather phenomena that occurred at the time of writing, so giving indications of changes up to a few thousand years ago in some places.

For earlier periods, evidence based on the study of natural systems that are dependent on climate can give us information for temperature or rainfall, or other climate variables, for particular times. Examples include vegetation, the type reflected in pollen found in a sediment core taken from a lake bed or the ocean floor. Fossils of plants and animals can serve the same purpose. Landforms can also become fossilized. A sand dune that is today stable thanks to a covering of vegetation was active some time previously, indicating that the area was once a sandy desert.

Deciphering the evidence for changing climates is a complex business, but considerable changes in desert climates in the geological past have been well documented for many of the world's arid regions. The Sahara is an interesting case in point. The extremely arid conditions of today have prevailed for only a few thousand years. There is lots of evidence to suggest that

6. Prehistoric rock paintings in the Tassili n'Ajjer mountain range in southeast Algeria indicate an environment that was much more verdant than the hyper-arid conditions that prevail in the central Sahara today

the Sahara was lush, nearly completely covered with grasses and shrubs, with many lakes that supported antelope, giraffe, elephant, hippopotamus, crocodile, and human populations in regions that today have almost no measurable precipitation. This 'African Humid Period' began around 15,000 years ago and came to an end around 10,000 years later.

Over rather longer timescales, up to around two million years ago, drastic changes occurred in climate all over the planet during major glacial cycles, or ice ages. The world's deserts were much larger during the ice ages because more of the planet's water was stored in ice and much less in the oceans than we see today. Smaller oceans and cooler global temperatures meant less evaporation and less rainfall, hence the larger deserts.

Globally, at the height of the most recent glacial period some 18,000 years ago, almost 50% of the land area between 30°N and 30°S was covered by two vast belts of sand, often called 'sand seas'. Today, about 10% of this area is covered by sand seas. One of these regions of once-active dunes is the Nebraska Sand Hills in the USA. The region is made up of 57,000 square kilometres of dunes, some up to 120 metres high, that are now stabilized by grasses and receive about 500 millimetres of rainfall each year. In central southern Africa, the remains of dunes that were probably active at a similar time are today covered by dense woodland and receive a mean annual precipitation of more than 1,200 millimetres. We have convincing evidence to suggest that similar large-scale shifts in climate belts have occurred throughout Earth's history, dating back hundreds of millions of years.

Some deserts, by contrast, have been dry for a very long time. Evidence from the geology and soils of the Atacama suggests to most scientists that it is the oldest desert on Earth, with a history of extremely arid conditions stretching back for 10 to 15 million years. Some consider that it has been continually dry for much longer, perhaps up to 200 million years.

Elements of climate such as water and wind operate on the geological fabric of the land to help shape the variety of desert landscapes we see today, as described in the next chapter.

Chapter 2
Desert landscapes

Desert landscapes can be spectacular and unique but also fantastically monotonous. In parts of the Mongolian Gobi, you can travel for many hours across distinctly unexciting level gravel plains. Just across the border with China, however, lie the world's largest sand dunes. Some tower over 400 metres in height, dominating a sandscape that is punctuated with more than 140 shallow lakes. Each lake is located in a depression right at the foot of a sheer precipice of sand that soars up to a razor-sharp sandy peak. The appearance of one of these lakes, glimmering like a pearl in the desert from behind the crest of a dune, is one of the most dramatic sights anywhere in the natural world.

Other remarkable desert landscapes are more widely known and attract millions of visitors every year. They include Uluru in the 'Red Centre' of Australia and the Grand Canyon in the southwestern USA. Fantastic landscapes can be found in any environment, of course, but deserts come with their geological fabric exposed. The sparse vegetation means that the structure of the terrain, its rocks, hills, and valleys, are laid out for inspection in a manner that is unmatched in most other parts of the world.

The ways in which these landscapes are moulded have fascinated generations of desert travellers and researchers. Rocks are broken down by a group of processes known as 'weathering'. The regular heating and cooling of rocks, effected by the large daily temperature ranges so typical of many deserts, causes minerals in the rocks to expand and contract. The frequent stresses and strains set up within a rock can eventually lead to fractures along lines of weakness, splitting the rock apart and breaking it down into fragments.

The presence of moisture, even in small amounts, enhances the efficiency of temperature extremes in fragmenting rocks. Similar effects also occur thanks to salt crystals that grow and prise rocks apart. In deserts where the temperature frequently oscillates above and below 0°C, any water present in rock crevices can have the same effect due to recurrent freezing and thawing.

Living organisms settle on the surfaces of rocks, in their cracks and pore spaces, often forming a very thin layer, or 'biofilm'. These microscopic organisms – fungi, lichens, bacteria, and algae – are thought to play an important part in the breakdown of rocks in deserts, exerting an influence through the acids they produce. Bacteria and lichens may also play a role in the formation of 'desert varnish', a dark brown to black coating commonly seen on exposed rocks. This rock coating occurs in many environments, but it is particularly noticeable in deserts. Effectively, it paints the topography with a glossy topcoat. Some desert researchers suggest that microbes affect the chemistry that fixes this paper-thin surface concentration of clay minerals, manganese, and iron.

The involvement of moisture in many of these forms of weathering in deserts may, at first, seem surprising given the very nature of the dryland realm. However, the previous chapter demonstrated that some water is frequently and sometimes regularly present in deserts, even if only in small amounts. River valleys and dry lake beds suggest that water is, or has been, present on occasion in much larger quantities as well.

Conversely, the terrain in some deserts clearly owes its origin to the work of the wind. Large areas covered by wind-blown sand are the most obvious example. The power of the wind to move huge quantities of fine material is also evident in the dense dust storms that can be traced on satellite images billowing out of deserts across large tracts of ocean. In many places, however, evidence can be found to imply the action of both wind and water in shaping the landscape. Interaction between the two forces is probably more common than either operating discretely, although lengthy periods may elapse between material being moved and laid down by one medium and it being moved again by the other.

At the largest scale, both wind and water operate to change details on a landscape that owes much of its fundamental character to its geological history. Uluru, in Central Australia, is one of just a few isolated hills in a landscape that is otherwise gently undulating. This is typical of so-called 'shield deserts', sections of an ancient super-continent called Gondwana which broke up between 100 and 200 million years ago. These areas have been worn down over many millions of years into vast plains, only occasionally broken by a rare hill in harder rock. Central Australia is a classic example, others include the Sahara, the Kalahari, and Arabia.

The predominantly even plains of the shield deserts stand in sharp contrast to the more dramatic scenery found in 'basin-and-range' deserts. These deserts are located in parts of the world with a more recent history of tectonic activity. Folding and faulting in the Earth's crust have thrust up mountains, flanked by intervening depressions and plains. This category includes the deserts of North and South America and many in Central Asia.

No two deserts look exactly alike. A huge wealth of detail is bound up in every terrain, each reflecting a unique geological, climatic, and ecological history. Indeed, the diversity of desert landscapes is probably as great as any on the planet.

Dunes

Sand dunes make up the most iconic of desert landscapes. Their sleek forms and fluid curves epitomize the stark natural beauty of a desert scene. Prominent among many desert pioneers to be tantalized by the rhythmic curves of the dune was Ralph Bagnold, founder of the British Army's Long Range Desert Group during World War II. Bagnold's observations in North Africa formed the basis of a seminal work on sand and dune movement.

Sand dunes come in a surprising diversity of sizes, ranging from small accumulations tens of centimetres in height and length to huge mounds measured in hundreds of metres. Shapes vary considerably too, dictated mainly by the amount of sand in an area and the pattern of winds.

In places where the supply of sand is limited on a hard desert surface and the wind blows predominantly from the same direction throughout the year, a dune called a barchan may form. This is an isolated dune with a crescentic, or half-moon, shape. The open horns of the crescent face downwind, and a barchan two or three metres high can move perhaps 15 metres in a year. Larger dunes move more slowly. The movement occurs with only minor changes in the dune's shape. Some studies have been able to identify the same barchans on series of aerial photographs and track their migration over several decades.

Most barchans range between 3 and 10 metres high, but larger examples have been recorded. Some of the barchans on the Skeleton coast of Namibia are 25 to 30 metres high and the Pur-Pur dune in Peru is 55 metres in height. Dune fields with hundreds of barchans, all more or less of the same size, are found in several desert areas, including the Pacific coast of Peru and the Atlantic coast of Mauritania.

The most common form of desert dune is a lengthy, slightly sinuous ridge known as a linear dune. Linear dunes range from a few hundred metres to nearly 200 kilometres in length and can reach heights of 200 to 300 metres. Good examples can be found in the Namib, the southwest Kalahari, parts of the Sahara, and in Australia's Simpson, Strzelecki, and Great Sandy deserts. In some places, linear dunes are aligned largely parallel to the prevailing wind direction. Australia's linear dunefields form a large anticlockwise whorl across more than one-third of the continent, in line with the dominant wind directions of a high-pressure cell.

Dunes can take on more complex shapes in regions where winds blow from numerous different directions. One of the most distinctive is the star dune, a peak of sand shaped like a pyramid with radiating sand ridges. These are found in parts of the Sahara, the Arabian Peninsula, and the Namib. In the Gran Desierto in Mexico, star dunes occur in linked chains. These Mexican examples reach heights of 150 metres, but star dunes exceed 300 metres in the Namib and in Algeria's Issaouane Erg. The highest star dunes, however, are more than 400 metres high: the highest dunes in the world. These are in the Badain Jaran in Chinese Inner Mongolia.

Most of the world's sand dunes are concentrated in vast sandy areas known as sand seas, or *ergs* (a North African Arabic word). The largest of these is the Rub al Khali, the Empty Quarter of Arabia, which covers about 600,000 square kilometres, an area larger than France. Several different types of dune are found in the Rub al Khali, including barchans, linear dunes, and star dunes. Other sizable *ergs* are found in North Africa and Central Asia. A distinct hierarchy of sand shapes are found in *ergs*, often producing a compound pattern. The largest dunes – so-called mega-dunes – are typically covered with smaller ones, and these smaller dunes themselves are covered with sand ripples.

The processes responsible for the accumulation of sand seas and the source of the sand are better known for some areas than

7. A star dune in the Rub al Khali, the Empty Quarter of Saudi Arabia. Star dunes dominate the eastern and southern regions of the world's largest *erg* or sand sea

others. The Ténéré desert, part of the Sahara in Niger and Chad, comprises two great sand seas, the Erg du Ténéré and the Grand Erg de Bilma. Sand that makes up these *ergs* has probably come from the foothills of the Tibesti Mountains in Chad. The material that makes up the Wahiba Sands, an *erg* in Oman, is attributed to the erosion of soft sediments on the Arabian Sea coast that were exposed during periods of lower sea levels. Some of the Wahiba's sand may also have been blown from the Rub al Khali.

Dune sand is most commonly quartz, although dunes can be composed of sand-sized grains of other material, including carbonates, clays, salts, or even ice. Quartz typically has a yellow colour, but in many deserts dunes take on an orangey-red hue as iron oxide is deposited on individual grains. The white sand dunes found in parts of the Chihuahuan desert – in northern Mexico's Cuatro Ciénegas basin and in New Mexico, USA – are so coloured because they are made up of pure gypsum, a mineral containing calcium.

The build-up of sand can continue for many thousands of years, and hence *ergs* and dunes may contain evidence of how climate has varied over time. One explanation of how the giant star dunes of the Badain Jaran have reached their enormous height suggests that alternate periods of relatively wet and dry conditions over tens of thousands of years have played a part. During wetter phases, vegetation grows on the dunes, helping to stabilize the shifting sands, a process reinforced by the formation of a crust on the dune surface. Calcium blown on to the dunes from the Badain Jaran's numerous lakes mixes with rainwater to develop a hard cement. This combination of vegetation and cementation fixes the dunes, which then become covered with loose sand once more during a drier period like the climate of today, hence gaining height incrementally over time.

Crusts of a different kind are common today on dunes elsewhere in China. Biological soil crusts dominated by lichen are very widespread in the Gurbantunggut desert in the Junggar Basin, where they play an important role in stabilizing sand dunes. It is not uncommon to see vegetation on dunes, particularly in many of the less arid deserts, and vegetation also helps to reduce the movement of sand in dunes. Physical geographers distinguish between active dunes that migrate; dormant dunes where activity is reduced for some reason, such as plant cover or a change in the supply of sediment; and 'relict' dunes that are very stable, often with a high proportion of vegetation cover.

Dunes are not only found in the world's deserts – a variety of dune shapes occur on beaches, on the seabed, in snow, and on the planet Mars – but they are certainly very distinctive features of many desert landscapes. Around one-third of the Arabian subcontinent is covered by sandy deserts, for example, and as much as 85% of the Taklimakan desert, in the Tarim Basin of northwestern China, consists of shifting dunes. In other deserts, dunes are much less common. There are no large *ergs* in any North or South American desert, where dunes cover less than 1% of the arid zone.

Yardangs

A distinctive landform found in many deserts is a streamlined
ridge that classically resembles an upturned boat with a tapered
profile. This is the yardang, a name derived from a Turkmen
word meaning ridge or steep bank, first introduced to a wider
audience by the Swedish desert explorer Sven Hedin. Hedin
became fascinated by the yardangs of Lop Nor during his travels
in Central Asia at the end of the 19th century.

The streamlined shape suggests the dominance of wind in
the formation of yardangs, but other processes may also
contribute, including sporadic incision by rivers, slumping
and sliding. Yardangs have been found at a considerable range
of scales, from ridges just a few centimetres high, through
features measuring metres in height and length, to so-called
'mega-yardangs' that may be tens of metres high and several
kilometres long.

Mega-yardangs occupy hyper-arid desert areas and are large
enough to be readily identified on satellite images as regional-
scale grooves in the landscape all aligned in the same direction.
This is further testament to the power of the wind because the
alignment is parallel to the prevailing wind direction. Those in
the western part of the Lut desert in Iran all run in a NNW–SSE
direction, corresponding exactly to the direction of the prevailing
'wind of 120 days', or *Bad-i-sad-o-bist roz Systan*, a fierce summer
wind that can reach hurricane force.

The 'wind of 120 days' is also well known for its violent sand
and dust storms. Thousands of these tiny particles bombarding
the rock walls of a yardang help to sculpt its shape, literally
sandblasting the landscape into streamlined ridges. The same
process is notorious for causing structural damage to buildings in
the region. Houses in Seistan, in neighbouring Afghanistan, are
built with dead walls facing the direction of the wind. The Lut

desert yardangs are up to 80 metres high and extend for tens of kilometres. They occupy an area about 150 kilometres long and 50 kilometres wide.

Such large yardangs may have been shaped over millions of years and are found in several other very dry deserts, including northern Saudi Arabia, the central Sahara, the Namib, and the Ica Valley region of central Peru. Similar features have been identified from satellite images on the surface of Mars, an arid planet where wind plays a very significant role in shaping the landscape.

Desert pavements

A wide-open expanse of generally flat, stony ground is another typical desert landscape thought to have been formed by the action of the wind. Desert pavements have a covering of small rock fragments and pebbles, sometimes coarse, sometimes rounded, usually only one or two stones thick, set on a base of much finer material.

Some of these pavements are very extensive. They cover large parts of the Gobi desert, for instance, and the name '*gobi*' is often used as a general term for desert pavements. These rather tedious landscapes have attracted a number of other regional names, including 'gibber plains' in Australia, '*hammada*' in North Africa, '*serir*' and '*reg*' in other parts of the Arab world. Similar landscapes are also found in many mountainous and polar regions, but they are particularly associated with hot deserts.

Frequently the stones in a desert pavement are covered in desert varnish. Sometimes the stones have also been smoothed by the abrasive action of wind-borne sand and dust, and may even be sculpted into a series of flat facets that meet at sharp angles. The overall shape is streamlined, with a narrow point on the windward side. These wind-sculpted stones are called 'ventifacts'.

8. A small area of desert pavement in the Sahara in Niger (Note the pen at top right of the photograph for scale)

The desert pavement is usually explained as an armoured surface left behind after the wind has blown away finer material between the stones. The result is the stony surface that protects the silt and sand below from further deflation. However, the windy origin may not be quite so straightforward. In many deserts, fine loose material is protected by a thin crust and the wind can only remove it if this sheath is first broken up, perhaps on a previous windy day, or by animal activities or vehicular traffic. It is also likely that the sorting of small grains by water has played a part in desert pavement formation. Sheet flooding after occasional rainstorms can provide more than enough running water to dislodge and remove fine material in a similar way to the wind.

Another theory suggests that the stony fragments may have been forced upwards through the fine material below to emerge at the surface. Laboratory experiments have shown that alternate freezing and thawing of mixed sediments can cause larger particles to migrate toward the ground surface, a process that may be particularly effective in the higher latitude deserts. Other

42

mechanisms that could force stones up to the surface include alternate wetting and drying of the soil and heaving caused by the formation of salt crystals.

Another theory still sees deposition, rather than erosion, by wind as a key factor. This explanation suggests that pavements form as dust from elsewhere settles on the ground and infiltrates below the stony surface. In doing so, the dust effectively pushes up the larger stones, keeping them on the surface.

Of course, any one or a combination of several of these explanations may apply in particular cases. Whatever their mode of formation, it seems that these stony surfaces may be very old. One of the few desert pavements to be dated, west of Lake Eyre in Central Australia, is thought to be at least two million years of age.

Rivers

It may seem paradoxical, but some of the world's largest rivers occur in deserts. The explanation is simple: very little of their water comes from the desert itself. Most is supplied from well-watered regions outside the dryland environment. These are so-called 'exotic' rivers.

The other major type of river found in deserts is one that arises within the desert itself, and these rivers differ from their exotic counterparts in several ways. One of the most striking differences is the most straightforward. Water commonly flows throughout the year in exotic rivers, but the flow of water in other desert rivers is highly variable, and for much of the time the river has no water flowing in it at all. One study of a river bed in the northern Negev desert showed that the channel contained water for just 2% of the time, or about 7 days per year. This type of usually dry river is called a *wadi* or *oued* in North Africa and the Middle East and an *arroyo* in Spain, the US southwest, and in Spanish-speaking parts of Latin America.

The classic example of an exotic desert river is the Nile, which rises in the Highlands of Ethiopia, Uganda, and Kenya and flows through the eastern portion of the Sahara before reaching the Mediterranean. Two of Central Asia's major rivers are also exotic: the Syrdarya and Amudarya both arise in the Parmir mountains and flow through the Kyzylkum and Karakum deserts. These examples also illustrate another two-fold division of desert rivers. The Nile flows to the ocean, but the Syrdarya and Amudarya do not. Much of the drainage in deserts is internal, as in Central Asia. Their rivers never reach the sea, but take water to interior basins. The Syrdarya and Amudarya deliver their water into the Aral Sea, a large inland lake. The perennial flow of water in exotic desert rivers provides a constant source of this vital resource, a great boon in an otherwise capricious environment. The Nile supported an ancient civilization in the desert in Egypt and continues to provide virtually all the water used in Egypt today. Great desert cities also arose on the Tigris/Euphrates, the Indus, and the Syrdarya.

Some desert countries, including Bahrain and Saudi Arabia, have no permanent rivers or streams whatsoever. The flow of water in many of the episodic rivers that arise in deserts is spasmodic and unpredictable. This is a simple reflection of desert rainfall being highly variable in time and space. When precipitation does occur, the short-term total amount of water that falls sometimes far exceeds long-term annual averages, and in the same way while the amount of water in desert rivers is low on average, at its peak a river carries very large amounts of water.

Cooper Creek, one of the major rivers of the Lake Eyre Basin in Australia, has one of the most unpredictable flow regimes of any river in the world of comparable size. Like most Australian desert rivers, for much of the time Cooper Creek is little more than a string of unconnected muddy waterholes. When water does flow, the creek becomes a mighty river. Floodwaters can inundate tens of thousands of square kilometres and may take many weeks before they stop flowing.

The large volumes of water involved during such flooding of desert rivers can build up quickly – so-called 'flash floods'. These very powerful events carry great loads of sediment. Such large floods do not happen very frequently, but they have a long-lasting effect on the shape of desert river channels. Flash flooding is also a serious hazard in the desert environment, often causing loss of life and great damage to roads, buildings, and other infrastructure.

Assessment of the power of desert rivers relative to rivers in other environments has thrown up the widely held belief that rivers in semi-arid zones are among the most effective at eroding the landscape. In wetter areas, more vegetation grows, and this increases protection of the soil surface and reduces the amount of water that runs off the soil to cause erosion; in drier areas, vegetation is sparser but less runoff is available to cause erosion. Although water can still be very powerful on the rare occasions it does flow, wind is a more pervasive agent of erosion in the driest desert areas.

Badlands

A terrain of skeletal ridges, riven by steep-sided gullies and devoid of vegetation, makes up arguably the desert's most distinctive scenery associated with water. These are visually dramatic landscapes known as badlands: barren soil surfaces that have been deeply dissected by rapid water erosion.

Badlands are particularly associated with the deserts of the US southwest. The name itself is probably derived from the French *mauvais terres à traverser* – land that is difficult to cross – because French pioneers were among the first Europeans to explore the Great Plains of western North America where extensive areas of badlands exist. Bleak, rugged, and difficult to access, these areas have little economic value, although their striking character has made some badlands important tourist destinations.

Although undoubtedly dramatic, badland areas make up only minor portions of deserts. These landscapes are found throughout the desert areas of western North America and other locations include the Zapotitlan Salinas drylands in Central Mexico, the Rif mountains in Morocco, the Kasserine area of central Tunisia, the Great Karoo in South Africa, the Tabernas desert in southeast Spain, and the Zin Valley in the northern Negev desert of Israel.

Badlands are also found in many other climatic regions, on a wide range of soils, and similar landscapes can be caused by poor land management in areas of intense cultivation, strip-mining, and on spoil heaps, but the lack of vegetation is a key factor in their development anywhere. Sporadic, intense rainfall causes rapid erosion across the landscape on steep slopes but also below the surface. Piping or tunnelling takes place as water seeps into cracks in the soil and enlarges them, creating elongated tunnels that can measure up to 30 metres long and 2 metres high. Should the roof of a tunnel collapse, a new gully is revealed at the surface.

Alluvial fans

Rivers and streams that emerge from mountainous areas onto adjacent plains spread out and lose their capacity to carry sediment. As a result, large amounts of rocks, stones, sand, silt, and clay are deposited across the landscape as an alluvial fan. These features have a semi-circular fan-like shape when viewed from above, although their shape in three dimensions is more like a cone. The apex of the fan is located at the point where the river emerges from the base of the mountain front.

Alluvial fans are found in many mountainous areas and arctic environments, but they tend to be bigger and best developed in deserts, particularly in basin-and-range topography. Individual fans can be large, in some cases more than 50 kilometres wide, and rise many hundreds of metres above the plain. They may also take a considerable length of time to form. The Milner Creek fan,

9. This satellite image over the Zagros mountains in southern Iran shows a large semi-circular alluvial fan that has developed at the point where an ephemeral stream exits steep slopes. Spreading out from the fan is a belt of agricultural fields that depend on groundwater stored within the fan

in the White Mountains of California, sits upon a desert surface that has been dated to some 700,000 years ago. Hence, the Milner Creek fan is nearly three-quarters of a million years in age and continues to grow.

Fans often coalesce to form an overlapping series that makes up an apron of alluvium known as a 'bajada'. Bajadas may also assume considerable dimensions. The alluvial fans south and west of the Oman mountains in southeast Arabia form a major, gently sloping bajada plain measuring about 500 kilometres by 200 kilometres.

The streams that form fans frequently change their course, so that at any particular time only some parts of a fan are active.

Sometimes the water flowing is so full of small sediment that it is better described as a mud flow, but these mud flows may also be carrying huge boulders, some the size of a house.

A substantial part of the floodwaters that occasionally run across a fan seeps into the fan itself, and over time this infiltration builds up a store of groundwater. This water has made such sites attractive for human settlement over the ages, and today many large desert cities, along with their associated infrastructure, are located on or beside alluvial fans. For instance, there has been a trading settlement on the site of today's Jordanian port town of Aqaba for about 3,000 years. Aqaba lies at the southern end of the hyper-arid Arava Valley which is flanked on both sides by steep mountain fronts. Aqaba is dominated by the large alluvial fan that emerges from Wadi Yutm. On the opposite side of the valley, across the border in Israel, the town of Eilat is located on a bajada slope created by several coalescing fans.

Buildings, roads, railways, and runways located on or near alluvial fans face the hazard of flash flooding and its associated erosion and deposition. Floods on alluvial fans are particularly dangerous because they travel at extremely high speeds and carry tremendous amounts of sediment and debris. It is also virtually impossible to predict either when or where one will next occur.

Lakes and wetlands

Areas that are continuously or periodically inundated by water occur in many deserts. This may at first appear contradictory, given that deserts are characterized by an overall deficit in surface water due to the low ratio between precipitation and potential evapotranspiration. None the less, local circumstances can have the effect of making the surface water balance positive for all or part of the year, hence the existence of permanent, seasonal, or ephemeral lakes and a number of different types of shallow wetland.

The water found in some of these aquatic features is supplied by exotic rivers flowing into areas of internal desert drainage. The Aral Sea in Central Asia has been mentioned in this chapter. An equivalent wetland area is the Okavango delta, in the Kalahari desert in northern Botswana. In spite of its semi-arid to arid setting, the Okavango delta receives enough water from the tropical highlands in Angola, brought by the Okavango River, to sustain about 4,000 square kilometres of permanent wetland. Seasonal flooding extends the area of the wetland to as much as 12,000 square kilometres. The Okavango is the only inland delta in sub-Saharan Africa. It is a complex of islands, ridges, floodplains, permanent and seasonal swamps, and a seasonal freshwater lake. These areas provide habitats for many species of bird and other wildlife not normally associated with a desert environment, including 68 species of fish and more than 1,000 different plant species. Among the larger mammals found in the Okavango are elephant, hippopotamus, wild dog, and the sitatunga, a swamp-dwelling antelope.

Elsewhere, permanent water bodies may be maintained by flows from stores of groundwater. This is how the Ounianga lakes of northern Chad have survived in the hyper-arid central Sahara. The four principal lakes are at Ounianga Kebir, midway between the Tibesti mountains and the Ennedi massif. They occupy wind-scoured depressions between a series of sand ridges and the groundwater that maintains them has not been recharged for thousands of years. Ounianga Kebir is a classic desert oasis, an isolated, fertile, green, watered spot surrounded by extremely arid desert. Although many of the lakes are saline, cattle herders take their livestock to them for water and grazing, and the oasis supports about 25,000 cultivated date palms.

The Ounianga lakes are located on the edge of what was once the largest lake in Africa, covering at least 350,000 square kilometres, an area greater than all the Great Lakes of North America combined. This was Palaeolake Megachad, probably some 170 metres deep about 10,000 years ago, its dimensions implied from prominent ancient shorelines that remain visible in the landscape today. Most of this vast ancient lake bed is now dry, with the exception in the southern part of the basin of modern-day Lake Chad, a much smaller body of water that has seen great variations in size over the last 50 years but covers less than one-tenth of the ancient lake's expanse.

The northern part of the former lake bed is called the Bodélé Depression, an area that straddles the border between Niger and Chad. The depression is covered by a crumbly, grey-white rock called 'diatomite', made up of the fossil remains of diatoms, single-celled aquatic plants. This material is very easily eroded and the Bodélé Depression is the world's largest source of wind-blown desert dust. Vast quantities of diatomite are blown from this area by the Harmattan wind across West Africa and over the Atlantic Ocean. Scientists can only estimate the amount, but it may be as much as half a billion tonnes a year.

Desert depressions, whether or not they contain water, are formed in a variety of ways. These include subsidence due to tectonic activity, scouring by the action of the wind, and excavation by large mammals. The Qattara Depression, in the eastern Sahara in Egypt, is at its lowest point 134 metres below sea level. The generally accepted explanation of its origin is by wind deflation to a base controlled by the level of underlying groundwater. However, other explanations suggested include incision by an ancient river and a salt-weathering origin. Extensive saline deposits cover the floor of the depression.

Evidence that wind has had at least some impact on the formation of many desert depressions comes in the form of crescent-shaped dunes that are commonly found on the downwind edge of depressions. These 'lunette' dunes are made up of material excavated from the surface and deposited by the wind.

Many of the lakes situated in desert depressions are salty to some degree. When dry, a hard saline crust often forms and this crust is frequently covered by a large polygonal pattern of slightly raised cracks that are formed as the salt dries. These saline desert lakes are frequently called 'playas', but numerous other names are also used, including *sabkha* in Arabia, *kavir* in Iran, *chott* in North Africa, *pan* in southern Africa, *salar* and *salina* in South America. The Salar de Uyuni, 3.6 kilometres above sea level on the Altiplano in Bolivia, is the largest salt flat on Earth. It is very flat indeed, exhibiting less than a metre of vertical relief over its 9,000 square kilometre area. Like many playas, Uyuni is usually flooded in the rainy season, from December to March in this part of the Altiplano.

The Salar de Uyuni is the lowest point of an internal drainage basin that has undergone many cycles of inundation and evaporation over the past 50,000 years. A similar situation has occurred in North America since the last ice age. The Great Basin desert in the southwestern USA is, despite its name, actually made up of many depressions, and around 100 of these basins contained lakes at that time. Lake Bonneville was the largest, with a depth of some 330 metres and an area of more than 50,000 square kilometres. Today's Great Salt Lake is a small remnant of that much greater body of water.

The water, minerals, and flat topography found in playas have been important to human populations since prehistoric times. They have become sites for urban development, industrial use, for runways and race tracks. Two of the four playas within the boundaries of Edwards Air Force Base, in the Mojave desert of southern California, are used as runways for military aircraft and the Space Shuttle. The Bonneville Salt Flats, another remnant of ancient Lake Bonneville, is frequently used for attempts on land-speed records. All of these activities are prone to the sporadic threat of flooding, as some residents of Salt Lake City found to

10. A view across Salar de Uyuni, the world's largest salt flat, in southern Bolivia. The brilliant white surface crust is almost pure halite or common table salt

their cost in the 1980s when the Great Salt Lake reached an historically high level.

Volcanoes and impact craters

Evidence of volcanic activity, both ancient and contemporary, is clear in several desert regions. Very old volcanoes form some of the Sahara's mountainous regions: the Hoggar mountains in southern Algeria and the Tibesti Massif of northern Chad. Similarly, the landscape comprising the Druze mountain foothills in Syria, the eastern Badia in Jordan, and adjacent parts of northern Saudi Arabia are covered by plains of basalt created by ancient flows of lava, punctuated by extinct volcanic cones. In Arabic, these lava fields are known as *harraat* (singular: *harrah*), and numerous *harraat* also cover some 180,000 square kilometres of western Saudi Arabia, together forming one of Earth's largest alkali basalt regions. One of the most recent lava flows in the area was in AD 1256, when two months of volcanic eruptions near Madinah sent lava flowing for 23 kilometres, threatening to inundate the town itself.

Current volcanic activity occurs in the Danakil desert, located on the Red Sea coast at the northern end of the East African rift valley in Ethiopia, Eritrea, and Djibouti. Hot springs and solidified lava flows surround Erta Ale, an active volcano in northern Ethiopia, which is part of the 90-kilometre-long Erta Ale volcanic range in the central part of the Danakil. Erta Ale volcano has very gentle slopes, rising to just over 600 metres above a base that spans roughly 30 kilometres. At its summit the volcano contains an active lava lake, one of just four in the world.

Volcanic activity has also commonly been a fundamental part of the tectonic upheavals that created the basin-and-range deserts. Signs of vulcanism punctuate the landscape of some of the North American deserts. Ancient lava flows and well-preserved volcanic cones can be found in parts of the eastern Mojave desert where 7 million years of eruptions ended around 10,000 years ago. Volcanic rocks associated with the uplift of the Andes also dominate the Altiplano-Puna Plateau in South America.

A number of circular, bowl-like features caused by the impact of meteorites have also been identified in desert areas. Some of the best examples are found in the ancient shield deserts, and are relatively easy to recognize in otherwise flat plains. At least 20 craters reported from the Sahara are thought to be of an impact origin. Others have been proposed following their identification on aerial photography and satellite imagery, but still await verification in the field. Some of the Saharan craters are considerable both in terms of size and age. The Aorounga impact crater in northern Chad, which is about 17 kilometres in diameter, was caused by an asteroid or comet that fell to Earth some 200 million years ago.

Most impact craters known on Earth are single features, but a few fields of craters, created by meteor showers, also exist, and some of these are in deserts. The Wabar crater field in southern Saudi Arabia, in the Rub al Khali desert, is one of the youngest anywhere. Its four relatively small craters, with diameters between 17 and 100 metres, are thought to have been created only about 300 years ago. Probably the largest impact crater field on Earth covers an area of more than 4,500 square kilometres close to the Gilf Kebir Plateau in the Sahara of southwestern Egypt. The field consists of 13 craters that range in diameter from 20 metres to 1 kilometre. Given the large area of the crater field, it was possibly created by several meteorites that broke up when entering the atmosphere. This is unusual since all the other known crater fields on Earth extend over an area less than 60 square kilometres and can all be explained by the break-up of a single meteorite.

Soils

Most deserts have no soils. Huge areas are simply made up of bare rock or sandy sediment that can hardly be called a soil. Soils tend to become more noticeable towards the desert fringe where more water means soil forms more easily.

This is not to say that no soils are found in any deserts, but specifying how much and where depends in part on how a soil is defined, which is not as straightforward as might be expected; it is easier to say what soil is not. Soil is the surface of the earth that is not rock, air, or water. However, most would agree that soil includes all of these elements – rock fragments, air, water – plus organic material, living or dead. More precise definitions depend on the interests and needs of those doing the defining. Inconsistencies between definitions can create confusion, especially at the limits of what is and what is not considered to be a soil.

None the less, the soils found in deserts are distinctive in a number of ways. They tend to form slowly, they are shallow, the texture of their material tends to be coarse, and they tend to hold substances that in other environments would be washed away. These characteristics all stem from the fact that there is little moisture in deserts.

Desert soils receive large amounts of airborne dust settling on them, which mixes with the base rock material that forms the soil. Their slow speed of formation also means that many desert soils retain features that reflect previous conditions. Examples include large areas of hard crusts, or 'duricrusts', that were formed under wetter conditions. These crusts include 'laterites' (which are rich in iron) and 'silcretes' (rich in silica) which are found in many deserts at or just below the earth surface. Both laterites and silcretes are widely distributed across the Australian landscape, for instance.

Salt is a common constituent of desert soils. The generally low levels of rainfall means that salts are seldom washed away through soils and therefore tend to accumulate in certain parts of the landscape. Large amounts of common salt (sodium chloride, or halite), which is very soluble in water, are found in some hyper-arid deserts. In deserts with rather more precipitation, the common salt tends to be washed out, or 'leached', from the upper soil to leave another salt, gypsum, at the surface. On desert fringes where both common salt and gypsum have been leached away, another form of duricrust (calcrete, which is rich in calcium carbonate) is common.

Where these salty accumulations occur depends on the minerals in local rocks, as well as other sources of salts, including the ocean and volcanoes. Calcrete is the most widespread type of desert duricrust. Crusts of halite commonly occur in playas (the Salar de Uyuni is halite), and gypsum is particularly common on the northern fringes of the Sahara, forming most of the chotts in Algeria and Tunisia. One characteristic form of gypsum crystal growth is the sand rose, an assemblage of crystals that are flat and shaped like blades, or the petals of a rose, thinning towards their edges.

Large quantities of many kinds of salts, some of them rather unusual, are found in the soils of the Atacama desert, where they have accumulated thanks to the long history of extremely arid conditions. The salts include halite and gypsum but also less common deposits of iodates, perchlorates, and nitrates. Nitrates can be found in other deserts, but the Atacama deposits are unique because of their high concentrations. They have been mined and exported as natural fertilizer since the 1830s (see Chapter 5).

Chapter 3
Nature in deserts

Deserts look barren, desolate, and lifeless for much of the time. In general terms, they have few plants and animals, but those that are found in deserts are remarkably diverse. A desert traveller may encounter camels or cacti, flamingos or frankincense trees, shrimps or stone plants, depending on where he or she is in the world. Indeed, despite their bleak reputation, deserts harbour a multitude of natural wonders, including perhaps the world's oldest living organism (a creosote bush in the Mojave that may be over 11,000 years old), its fastest rodent (the mara, or Patagonian hare, that can reach 45 kilometre/hour), and the only reptile known to have an annual life cycle (the Labord's chameleon in the semi-arid southwest of Madagascar).

The previous chapter looked at the physical fabric of desert topography, but its wildlife, the living content, forms an equally integral part of any landscape. Deserts contain many extraordinary ecosystems with their own unique assortments of flora and fauna. From the fog oases of the Atacama to the dragon's blood groves of Socotra, from Western Australia's mulga shrublands to South Africa's Namaqualand termite mounds, the nature of deserts can rival that found in any other environment.

None the less, deserts certainly present some serious challenges to the survival of plants and animals. The very beginnings of life originated in water and living things then slowly invaded dry land. This gradual, evolutionary process meant organisms having to cope with radically different physical environments. Greatest among these differences are the lack of water, highly variable temperatures, and an excess of shortwave radiation, or sunshine. These contrasts are most extreme in the world's deserts.

One way of assessing living things is to measure the amount of organic matter or biomass produced over an average year, principally through the process of photosynthesis. This is called the 'net primary production', and comparing the amount produced in deserts to that of other environments is revealing. Deserts are among the planet's least productive ecosystems, up to 30 times less productive than tropical rainforests and wetlands.

This is not to say that deserts are dead – far from it, in fact. Many deserts are very rich in rare and unique species thanks to their evolution in relative geographical isolation. Many of these plants and animals have adapted in remarkable ways to deal with the aridity and extremes of temperature. Indeed, some of these adaptations contribute to the apparent lifelessness of deserts simply because a good way to avoid some of the harsh conditions is to hide. Some small creatures spend hot days burrowed beneath the soil surface. In a similar way, certain desert plants spend most of the year and much of their lives dormant, as seeds waiting for the right conditions, brought on by a burst of rainfall. Given that desert rainstorms can be very variable in time and in space, many activities in the desert ecosystem occur only sporadically, as pulses of activity driven by the occasional cloudburst.

When a rainstorm does occur in a desert, the landscape is rapidly transformed from a sparse and desiccated array of skeletal vegetation to an explosion of multicoloured flowers on plants that are designed to complete their life cycles rapidly. Short-lived or ephemeral ponds and other water bodies also briefly come alive with aquatic creatures that have survived the prolonged period without water perhaps as eggs (as do shrimps) or by burrowing deep beneath the ground in moist holes (like some types of frog).

The general scarcity of water is the most important, though by no means the only, environmental challenge faced by desert organisms. Limited supplies of food and nutrients, friable soils, high levels of solar radiation, high daytime temperatures, and the large diurnal temperature range are other challenges posed by desert conditions. These conditions are not always distributed evenly across a desert landscape, and the existence of more benign microenvironments is particularly important for desert plants and animals. Patches of terrain that are more biologically productive than their surroundings occur in even the most arid desert, geographical patterns caused by many factors, not only the simple availability of water.

Microenvironments present more favourable opportunities for some wildlife, but in general terms the environment in all deserts poses strong challenges for survival, and these challenges have often been met by the evolution of similar external characteristics in plants and animals. This so-called 'convergent evolution' describes the way in which species that are unrelated or distantly related, often in widely different parts of the world, have acquired the same biological traits in response to the same challenge. Hence, some plants from the cactus family (Cactaceae), the euphorbia family (Euphorbiaceae), and the milkweed family (Asclepiadaceae) look very much alike because they have adapted to desert conditions in essentially the same ways: all are 'succulents', that is plants with swollen or fleshy tissues designed to store water; all have developed fleshy, column-shaped stems (providing the water-storage capacity), reduced leaves (limiting transpiration and water loss), and spines (protecting the leaves from animals that might eat them). Yet these are three fundamentally different families of flowering plants. The cacti are native only to the Americas (with one exception among some 2,000 species). Although some cacti have been introduced to other parts of the world, desert plants elsewhere that look like cacti to the untrained eye are not cacti at all. The apparently similar members of the euphorbia and milkweed families occur mainly in the deserts of Asia and Africa, where the ecological role they play is analogous to that of the cacti in the North and South American deserts.

Convergent evolution also occurs in the animal world and some good examples are found in desert lizards. An Australian lizard, the thorny devil (*Moloch horridus*), is strikingly similar in several ways to the desert horned lizard (*Phrynosoma platyrhinos*) found in North American deserts. Both are covered by sharp, protective spines; both are camouflaged by brown and sandy body colours; and both eat ants. The two lizard species are only distantly related, but they share much greater similarities than either shares with its closest living relatives.

In spite of these intriguing examples of evolutionary convergence, there remains an extraordinarily wide variety of desert wildlife, a range that matches the great variety of rainfall patterns, temperature regimes, topographies, and environmental histories that make up the dryland realm. Each desert has generated its unique organisms, often fantastic life forms that add up to a diversity of adaptations virtually unrivalled elsewhere in the natural world. Bizarre growth forms, surprising behaviours, and barely credible resilience combine to make the desert a place of matchless biodiversity as well as immense natural splendour.

Plants

Plants have adapted to live in all the numerous combinations of conditions found on our planet, and they have done so by developing a wide range of strategies to cope with the situations they face. The numerous ways plants have developed to survive the stresses of desert conditions are classified simply into a spectrum with those designed to tolerate stress at one end and those designed to avoid stress at the other.

Plants that tolerate stress are generally more efficient at using what sparse resources are permanently available in the desert. Many of the adaptations designed to avoid stress also combine with an ability to make the most of resources that become plentiful for short periods only, a pulse of abundance typically driven by a sporadic rainstorm.

The limited supply of water, high temperatures, and intense sunlight are the most serious stresses faced by desert plants. A wide range of adaptations have evolved in response to aridity, and these adaptations can be divided into those concerned with the physical shape and structure of the plant – 'morphological' adaptations – and others concerned with the plant's internal chemistry – 'physiological' adaptations.

The ability to procure and store water is essential in desert conditions. An extensive root system, either spread over a large area or penetrating deep into the ground, is typical of many desert plants. A shallow root system is designed to collect moisture from any rainfall or dew over a wide area, often many metres in all directions from the main body of the plant. Deeply rooted species tap into stores of groundwater which will see them through dry periods. The jand, or kandi tree (*Prosopis cineraria*), common in the Thar desert and in Oman, has a root system that reaches depths of 50 metres and more.

Morphological adaptations designed to maximize low levels of precipitation also occur above ground. The iconic dragon's blood tree of Socotra has an inverted umbrella-shaped canopy that intercepts moisture from mist, rainfall, and dew and channels it down its branches to be concentrated at the base of its trunk. Large succulent leaf rosettes are also well designed to harvest low-intensity rains and fogs, channelling the water to the roots. Scrub with a typical rosette shape is the dominant form of vegetation on mountainous slopes in Mexican deserts, where clouds and fog provide the most important source of water. Desert rosettes are also one of the most common fog-harvesting growth forms in the foggy coastal desert of the Atacama.

Most species of cactus have extensive, shallow systems of fine roots, often equipped with tiny hairs. Cacti store water in the succulent tissue of their stems, which have evolved to conserve water in a number of different ways. Relative to plants in most other environments, cacti have a reduced surface area through which water may be lost. Cactus spines are actually modified leaves which hinder airflow round the plant, so reducing evaporation, and help to reflect heat, as well as dissuading animals from feeding on its succulent flesh. Cacti have tough, waxy outer skins which also help to reduce water loss.

Stomata, the specialized pores found on the surface of plants that are essential to the process of photosynthesis, allow plants to take in carbon dioxide but also to lose water in the process. Desert plants therefore have to maintain a delicate balance between procuring the carbon dioxide needed for their survival and losing water via transpiration. This water loss is cut down in cacti simply because the number of stomatal pores is small. Cacti stomata are also sunken in tiny pits, a characteristic that helps to reduce transpiration further. The cost to the cactus of these mechanisms is to reduce the speed of growth.

Like cacti, many other species of desert plant have fewer stomata than plants established in other environments, but there are exceptions. One is the *Welwitschia mirabilis*, a plant found only in isolated communities in the Namib desert from central Namibia to southern Angola. The welwitschia has a particularly high density of stomata on its leaves because it procures much of its water from dew and fog directly through its pores. This very specialized plant commonly grows along dry watercourses and

11. Cacti are very well adapted to dry conditions. A columnar cactus like this one in the Chihuahuan desert, northern Mexico, has a shallow root system that typically extends laterally for 20 metres or more

also has a long taproot, allowing it to reach additional sources of water underground.

A physiological mechanism to reduce the loss of water by plants is a form of photosynthesis known as Crassulacean Acid Metabolism (or simply CAM), named after the family of succulents, Crassulaceae, in which the process was first described. In CAM photosynthesis, a plant takes in carbon dioxide through its open stomata at night when temperatures are lower, allowing the stomata to remain shut during the daytime, so minimizing water loss. The carbon dioxide is then stored inside the plant until the Sun rises and photosynthesis can begin. This form of photosynthesis, which is especially common in plants adapted to arid conditions, requires a great deal of energy, so those species that employ the CAM mechanism tend to be very slow-growing. Cacti are CAM photosynthesizers.

An extreme way in which some plants can tolerate a shortage of water is a strategy adopted by several species that are together commonly called 'resurrection' plants. The body structures of these plants are allowed to dry out and enter a dormant state during extended drought periods and can reabsorb water and resume their functions when moisture is once more available. The shrivelled, blackened appearance of these species, which have a crumbly texture while dormant, makes the plant seem to be completely lifeless. The return of normal functions with sufficient moisture makes the common name seem appropriate, as the plants seem to rise from the dead. The resurrection plants are most commonly found in the semi-arid parts of southern Africa, Australia, Brazil, and North America where drought is a frequent occurrence. They include many mosses which can remain in their dormant state for several years if necessary.

Another indispensable facet of plant life in most deserts is an ability to tolerate high temperatures and high levels of sunshine. Many plants maintain a relatively constant temperature using some of the mechanisms outlined above. Certain species actually move their leaves as the Sun angle changes, so minimizing the build-up of heat during the course of the day. The creosote bush (*Larrea tridentata*) is a case in point, its ability to orient its leaves in a north–south direction having engendered one of its common names: the compass plant. Another response to excessive sunshine is to grow largely underground, as happens with the 'stone plants' (*Lithops* spp.) endemic to the deserts of southern Africa. Other species respond by adjusting to seasonal temperature changes. This they do through the production of enzymes that can tolerate and operate at different levels of temperature.

Very high temperatures represent the main hazard to plants from fire, and numerous dryland plants have adapted to withstand them. In Australia, many desert eucalypts are fire-adapted, with bark that is thick and insulating or smooth, white, and heat-reflecting. Growth points beneath the bark, in branches or at the base of the trunk, allow fresh growth almost immediately after a fire. The above-ground parts of most shrubs are killed by severe fires, but many have adaptations allowing the species to grow again, including below-ground parts such as rhizomes and seeds that are fire-germinated.

Plants that avoid rather than tolerate stress typically enter dormancy when temperatures or moisture levels reach certain threshold levels. Many perennial desert species simply shed their leaves during times when little water is available. These species experience high rates of photosynthesis and growth during favourable conditions, but these processes slow considerably during less favourable periods. Some species, such as the ocotillo (*Fouquieria splendens*) of the North American deserts, can produce and then lose their foliage as many as seven times a year.

An adaptation found in many species of desert plants from all over the world is to survive unfavourable periods in the form of seeds, and to complete the life cycle during more favourable times. These are the annuals which go through rapid growth and reproduction during a wet period, often completing their life cycle in six weeks or less. In deserts with a discernible rainy season, this occurs in most years. In more arid deserts, the so-called 'annuals' bloom less frequently. In either case, by the time the water supply has gone, these plants have already created seeds, the form in which they will survive throughout the dry period until the next rains.

The seeds of annuals germinate only when enough water is available to support the entire life cycle. Germinating after just a brief shower could be fatal, so mechanisms have developed for seeds to respond solely when sufficient water is available. Seeds germinate only when their protective seed coats have been broken down, allowing water to enter the seed and growth to begin. The seed coats of many desert species contain chemicals that repel water. These compounds are washed away by large amounts of water, but a short shower will not generate enough to remove all the water-repelling chemicals. Other species have very thick seed coats that are gradually worn away physically by abrasion as moving water knocks the seeds against stones and pebbles.

In a wider sense, annual plants play a critical role in the ecology of deserts because the innumerable seeds left behind by these short-lived, ephemeral plants provide an immensely important food source for many creatures, particularly rodents, ants, and birds. Seeds provide a concentrated source of energy and nutrients but also a supply of water to an animal when the fats in seeds are metabolized inside its body. The population sizes of these seed-eaters, or 'granivores', are closely linked to the numbers of seeds available.

The annual flowering plants that exist for lengthy periods only as seeds are one of several groups of desert plants that may not be immediately obvious to the desert traveller. Desert shrubs, succulents, and small trees are all relatively large and easy to see, but deserts also harbour a wide range of small and microscopic plant life including liverworts, algae, lichens, fungi, and bacteria.

Indeed, in some extremely dry environments that can appear to be completely devoid of vegetation, these micro-organisms may be the only plant life present over vast areas.

Lichens are widespread in many deserts. They have no root system, absorbing water vapour from the atmosphere, and are therefore particularly extensive in the world's coastal foggy deserts. Lichens are a unique group of life forms that consist of two closely related parts, a fungus and a partner that can produce food from sunlight. This partner is usually either an alga, or occasionally a blue-green bacterium known as 'cyanobacteria'. Algal cells are protected by surrounding fungus which takes nutrition from the algae. When cyanobacteria are involved, nitrogen fixation is an additional benefit.

Lichens are particularly common on a wide range of rock types but they also occur on soils. In combination with other microscopic organisms, they often form a thin crustal layer on both rocks and soil surfaces in places that are unable to support higher plants. These 'biological soil crusts' are found in deserts all over the world. They are well adapted to arid environments, due to their extraordinary ability to survive periods of desiccation and temperatures up to 70°C, as well as high levels of sunshine, salinity, and alkaline conditions.

Less conspicuous lichens are those that live under translucent stones or actually inside translucent rock, where they inhabit microscopic crevices and pores. These astonishing habitats are like tiny natural greenhouses: enough sunlight penetrates to allow photosynthesis, and the microenvironment under stones or inside rocks traps the moisture necessary for life while providing protection against evaporation. Stony desert pavements composed of almost transparent rocks and minerals such as quartz, gypsum, halite, and sandstone are important sites for these microbial habitats. There are also some desert lichens that exist underground. Among these are several varieties of edible desert truffle found in drylands extending from Morocco eastward to Iraq and the Arabian Peninsula where they usually grow seasonally, with the onset of rains in springtime.

Animals

Desert animals, like plants, come in a wide range of shapes and sizes with a considerable variety of adaptations developed to enhance their chances of survival in desert conditions. These adaptations can be split into morphological and physiological features, as in the plant kingdom, but the mobility of animals adds an additional array of behavioural adaptations. Like plants, many animal species exhibit more than one type of adaptation to enable them to persist in the desert environment.

A common method for dealing with limited supplies of food and nutrients is to collect and store food during periods of abundance so that supplies are available during times of scarcity. In many deserts, rodents, ants, and other granivores collect and store seeds for this purpose. Increasing reserves of body fat is, in effect, another method of storage that is used by a variety of lizards in several world deserts as well as by the camel in its hump.

Becoming dormant for a period, lowering the metabolic rate and so using less energy, is an additional way to deal with shortages of food as well as a lack of water and low temperatures. Several types of desert organism spend most of their lives avoiding the lack of moisture in a dormant state and will become active only when sufficient water is available. Large numbers of protozoa, common single-celled aquatic creatures, are present in pools that form after a desert rainstorm but survive in the soil after the pool has evaporated by secreting a tough protective coating.

Some larger creatures behave in a not dissimilar way. The inland crab (*Holthuisiana transversa*) found in Australian low-lying desert areas survives long droughts by sealing itself in deep clay burrows where moisture levels are maintained for long periods. Several of the burrowing frogs that inhabit the arid areas of Australia shed flakes of skin to form their own cocoon, which provides an additional barrier against water loss by evaporation.

Numerous desert organisms survive lengthy periods of water deficiency in a resistant state, as an egg that will hatch when enough water is available for the creature to grow rapidly to maturity and reproduce, leaving large numbers of resistant eggs that persist during the next adverse period. These ephemeral species, the animal world's equivalents to the annual plant species, have a short life. They include certain aquatic crustaceans such as shrimps and many insects, including locusts.

The first reptile known to have this sort of annual life cycle was discovered in 2008. Labord's chameleon (*Furcifer labordi*) is found only on the island of Madagascar, in the semi-arid southwest. After emerging from its egg, the species has a life span of just four to five months. These chameleons all hatch at the onset of the rainy season in November and reach sexual maturity in less than two months. Reproduction in January and February is followed by death for all adults. The eggs persist for eight or nine months through the dry season before hatching with the arrival of the rains.

Another strategy to avoid harsh desert conditions is simply to leave the area. This is an option for larger creatures, flying insects, and birds that cannot withstand desert conditions in summer but use them in winter to obtain food and then migrate when not as much food is available. For less mobile animals, avoidance over much shorter periods occurs during the daytime when temperatures are high. Many spend the daylight hours in relatively moist underground retreats, emerging only at night when conditions are cooler and water loss is reduced.

The kangaroo rats (*Dipodomys* sp.) found in the North American deserts exhibit many of the adaptations for dealing with the hot desert common to small mammals living permanently in these environments. They are nocturnal, spending the daytime hidden in burrows that remain cooler and more humid than conditions above ground. Their physiology is well adapted to conserving water by producing highly concentrated urine and relatively dry faeces and, like many granivores, the kangaroo rats survive on a diet of seeds without drinking. Other strategies adopted by the kangaroo rats are more specialized. They supplement their supply of water from the metabolism of seeds by eating young green vegetation when available in the spring, and especially by hydrating dry seeds before consumption by storing them in humid burrows for a period. Water loss by evaporation is also kept to a minimum through a system of membranes in their nostrils that condense, collect, and recycle moisture. This adaptation also occurs in non-desert mammals, but the kangaroo rats need about 25% less oxygen than other mammals, meaning that they breathe less frequently, thus reducing evaporative water loss.

The need to conserve water is important to all creatures that live in hot deserts, but for mammals it is particularly crucial. In all environments mammals typically maintain a core body temperature of around 37–38°C, and those inhabiting most non-desert regions face the challenge of keeping their body temperature above the temperature of their environmental surrounds. In hot deserts, where environmental temperatures substantially exceed the body temperature on a regular basis, mammals face the reverse challenge. The only mechanism that will move heat out of an animal's body against a temperature gradient is the evaporation of water, so maintenance of the core body temperature requires use of the resource that is by definition scarce in drylands. This need encapsulates a fundamental difficulty for mammals that occupy hot deserts.

A few large desert mammals are exceptions to this general rule, however. Some are able to allow their body temperature to rise during the day, effectively storing heat, and release it at night when the air temperature falls below body temperature, so reducing the need to lose heat during the daytime by evaporative cooling (sweating and panting). Several species of gazelle, eland, and oryx have this facility, as does the dromedary camel (*Camelus dromedarius*), which is able to tolerate core body temperature fluctuations of up to 8°C (between 34°C and 42°C), a range that would be lethal to most mammals. The camel's long legs are an asset in the nocturnal cooling process, their considerable blood flow helping to dissipate the heat.

The one-humped dromedary camel has a number of other adaptations that make it extremely successful in desert conditions. Its ability to go without food and water for days is legendary. A camel can withstand severe dehydration (losing 20–30% of body mass) without ill effects, a level of resilience complemented by the facility to drink huge volumes of water rapidly. A 600-kilogram camel can drink up to one-third of its body mass in water in just three minutes, in one visit to a water source. The kidney of the camel is also designed to minimize water loss. Its urine is highly concentrated – when a camel is dehydrated its kidneys can concentrate the salts in its urine to twice the level of seawater – and its faeces are very dry.

The camel's thick coat of light-coloured hair reflects some of the sun's heat and insulates the skin tissue beneath, keeping it relatively cool. The effects of the fur, combined with sweating at the skin's surface, results in the skin on a camel's back being 30°C cooler than the surface of the fur.

The camel's ability to go for long periods without food is facilitated by its fat-filled hump. When food is in short supply, energy is produced by metabolizing fat, and the hump gets smaller. The dromedary, found primarily in the Sahara but also in many other hot deserts of Arabia, Asia, and Australia, has one hump. The Bactrian camel (*Camelus bactrianus*), which is more common in the colder deserts of Afghanistan, Central Asia, northern China, and Mongolia, has two humps for obvious reasons, given its need to survive very harsh winters.

Camels also have large padded feet that are well suited to walking on soft sand, translucent eyelids that allow them to see relatively well with their eyes shut, cutting down on intense solar radiation and problems from sand and dust in the air. Nostrils can also be closed to keep out dust.

A number of species are adapted to live on and in loose surface materials, typically sand, that present difficulties for movement and for burrowing. Morphological adaptations to crossing sand like the camel's wide, pad-like feet also appear on some types of lizard that have enlarged scales on their toes to dissipate their weight over a greater area, improving contact and ease of movement. Other creatures have developed specialized forms of movement, such as the side-winding locomotion practised by the rattlesnakes of North American deserts and sand-dwelling vipers in North Africa. The difficulties of burrowing into loose sand mean that many species effectively swim through it.

Sandy surfaces occur in many deserts, but a number of other desert terrains also present particular challenges to animals. One creature that is well adapted to living in highly saline environments is the red vizcacha rat (*Tympanoctomys barrerae*) of the Monte desert in western Argentina. This rat lives beside salt flats and eats vegetation that inevitably has a high salt content. It does so by stripping away and discarding the particularly salt-rich outer layers of the leaves with specialized incisors before consuming the less salty leaf interior. The rodent also has a unique bundle of bristles on the side of its mouth that assists in this process.

Another animal that seeks out saline environments in many deserts is the flamingo. These conspicuous wading birds feed on algae and crustaceans they filter from the shallow waters of

12. **Makgadikgadi salt pan in Botswana is one of the most important flamingo breeding sites in southern Africa**

playas. Some flamingo species inhabit high-altitude salars on the Altiplano-Puna Plateau of the Andes, while others are a common sight on the salt lakes of lower-altitude South American deserts in Argentina, as well as Arabia, eastern and southern Africa.

Some of the fish that occur in desert habitats have become adapted to live in salt water. One unusual and well-studied group of desert fishes is the desert pupfish found in springs, sinkholes, and marshes in North American deserts. Some pupfish species live in very extreme conditions, able to tolerate water salinities up to four times that of seawater and water temperatures of 45°C. Pupfish have also been found in water almost devoid of oxygen. These species have enlarged gills but also gulp air at the water's surface to supplement their meagre intake of oxygen from the water.

Obtaining fresh water in deserts is often a serious challenge for terrestrial creatures and several animals have adaptations to make the most of the limited resources available. In hyper-arid areas where rainfall is extremely rare and unpredictable, the occurrence of fog and dew plays an important role in the water economy of many organisms. Several creatures found in the Namib have developed interesting ways of 'fog-basking'. One is the sidewinder adder (*Bitis peringueyi*) which flattens its body against a cool sand surface during foggy periods, so increasing its surface area exposed to water deposition. The snake licks the water droplets off its body, raising its head to swallow using gravity.

Two species of tenebrionid beetles have also developed unique fog-basking behaviour in the Namib. This involves climbing up to the crest of a fog-swept dune at night and assuming a head-down stance, allowing fog to be deposited on their carapaces. The beetles drink drops of moisture that run down towards the mouth, gaining an average of 12% of body weight.

Dew is an important source of water for many desert animals, particularly those that feed on vegetation in the early morning. For most desert snails, dormant much of the year, a heavy dew provides sufficient moisture for them to become active, with high rates of water intake. An example is the snail *Trochoidea seetzenii*, which is widely distributed in the Negev. The Australian tarantula, *Selenocosmia stirlingi*, which occurs throughout arid areas in the centre of the continent, traps dew for drinking on a low silk-covered mound at the entrance to its burrow.

Size is a serious issue for desert invertebrates – those animals without a backbone – because these creatures lose water through their surface, and the smaller the organism the larger its surface area relative to its volume. Insects, however, have a waxy, waterproof outer layer to combat this problem. A covering of wax also makes scorpions better adapted to high temperatures and arid conditions than most other animals, although scorpions are not restricted to deserts. A scorpion's rate of water loss has been measured at just one part per 10,000 of its body moisture, the

13. Scorpions are remarkably resilient creatures, well adapted to the rigours of desert life. Some have been known to survive without food for more than a year, thanks to very low metabolic rates and low levels of activity

lowest recorded for any animal. Scorpions are also remarkably tolerant of dehydration, being able to survive losing up to 40% of their body fluid. Despite these capabilities, scorpions typically spend the daylight hours in burrows or seek some form of cover, and restrict their predatory surface activities to the night-time. This combination of behavioural traits is probably the most important adaptive mechanism for scorpions inhabiting desert areas, as it is for so many other creatures.

Wildlife patterns in the landscape

On their margins, where deserts fade imperceptibly into semi-deserts, the terrain tends to become dominated by grasses and their associated animal life. In the tropics, these are the savanna grasslands with scattered individual trees. Trees and large shrubs are almost completely absent, by contrast, in temperate grasslands: the Eurasian steppes, the veldts of South Africa, the pampas of Argentina, and the plains and prairies of central North America.

The distribution of species across these desert fringe landscapes is relatively continuous, a pattern seemingly only repeated in drylands where vegetation cover approaches zero in the most hyper-arid environments. Even in the driest desert, however, some patches of terrain are more biologically productive than others. These geographical patterns are a function of many factors including, but not limited to, the availability of water.

At a very local scale, moisture from dew tends to be concentrated beneath rocks and stones, trickling down impermeable surfaces and remaining longer in sheltered microenvironments where it supports the growth of algae and small communities of invertebrates and micro-organisms. Over slightly greater areas, large isolated trees are considered key organisms in arid and semi-arid environments. They create their own microclimate by providing shade that reduces extremes of temperature and limits excessive sunlight. The moisture content and fertility of the soil beneath the tree also tends to be higher, and these factors combine to create a microenvironment that improves the chances of seedlings germinating and surviving. Trees also become focal points for animal activity because they supply shade, scarce food resources, protection against predators, and, for birds, suitable sites for nests. These functions are provided in most of the Negev desert, for example, by its three native *Acacia* species. A similar role in the Sonoran desert is served by the ironwood tree (*Olneya tesota*).

Over larger areas, the type and density of vegetation in a desert landscape varies according to the nature of the terrain. Deep-rooted trees and other plants are usually more concentrated along ephemeral water courses because, although dry for much of the time, these channels typically maintain reserves of moisture below the surface. This vegetation in turn provides food and shelter for a variety of other organisms, so that a distinctive longitudinal ecosystem develops. Variations in other topographical factors, including aspect, slope, altitude, and salinity, may also encourage some plant types and discourage others. Major differences in desert vegetation can also be identified between, for instance, rocky hills and sandy plains.

Spots where a permanent source of water exists in an otherwise arid scene are places of markedly different plant and animal life. Foremost among these types of environment are the classic desert oasis maintained by groundwater and the corridors of lush growth found along rivers that flow through desert terrain. Given the fact that these areas are well watered and relatively green, in one sense their flora and fauna may not be particularly typical of deserts. Many of these species could not usually survive under desert conditions. From another viewpoint, however, the oasis and the fertile watercourse are as much a part of the desert scene as the camel and the cactus.

Cuatro Ciénegas basin in the Chihuahuan desert in Mexico contains an archetypal oasis. Translated from the Spanish, the name means 'four marshes', symbolizing an area of major springs that feed rivers, ponds, lakes, and marshland covering some 1,200 square kilometres. The basin's long isolation has given rise to numerous unusual forms of animal, so that 8 of the 16 species of fish in the basin are found nowhere else in the world. Other endemic species include crustaceans, molluscs, and turtles. Notable among its plants is a unique species of Mimosa (*M. unipinnata*), a shrubby tree about a metre in height.

Many studies of desert ecosystems have recognized a distinct patchiness in plant cover over large areas that is not related to obvious sources of water. These patterns are typically arranged as patches of relatively dense vegetation on a landscape otherwise characterized by low plant cover. In some cases, these dense patches may develop in time from the 'nurse' effects of single trees,

14. The dragon's blood tree, endemic to the island of Socotra in the Arabian Sea, has a canopy that channels precipitation down its branches to the base of its trunk. The microenvironment beneath the tree improves the chances of seedling survival, creating a small grove like this one

producing small groves of dragon's blood trees in Socotra, for example.

Elsewhere, the vegetation of these patches is more typically made up of smaller shrubs rather than trees. When viewed from an aircraft or on a map, two major types of pattern have been identified: banded and spotted vegetation. The banded pattern is known in Sahelian West Africa as *brousse tigrée* (tiger bush) because of its similarity to the stripes of a tiger. Following a similar analogy, the spotted pattern is sometimes called leopard vegetation.

The two patch patterns probably originate from common mechanisms. In both cases, the relatively lush areas reflect higher concentrations of vital resources: soil, soil moisture, and plant debris that is rich in nutrients. These resources become concentrated due to the effects of erosion moving material around the landscape. The tiger bush pattern is thought to result in areas where flowing water is the dominant process of erosion, and leopard vegetation occurs where wind is the major form of redistribution.

In the water erosion case, rainfall runs off bare soil but accumulates around any slight impediment to the flow such as a fallen log or subtle change in topography. In these spots, more water soaks into the ground, and when a sufficient amount of water is involved this allows significant plant growth to be maintained. This effect is reinforced over time, a 'positive feedback' process, creating the banded landscape pattern.

Brousse tigrée has been studied all across the Sahel, from Mauritania in the west to Sudan and Somalia to the east, as well as in Australia and North America. Spotted or leopard vegetation has been reported from several deserts in North and South America. These landscapes tend to be stable over time, and several investigations suggest that this redistribution of resources creates a more productive landscape overall than if resources were spread uniformly everywhere.

Vegetation patches also occur in deserts where fog is the most regular and frequent source of moisture, albeit in small amounts. The coastline of the Atacama is dotted with numerous isolated pockets of vegetation that survive solely on moisture from the fog. These vegetation patches, called 'lomas', are separated from each other by hyper-arid habitat where virtually no vegetation exists. In effect, they are terrestrial islands, or fog oases. Many of the plant species found in the lomas are so finely adapted to the fogwater that they are found nowhere else on Earth. At least 15 species of *Tillandsia*, or air plants, are the quintessential lomas vegetation type.

In other areas of the world, the pattern of dryland plant cover has different origins. Localized patches of richer soils in the Namaqualand desert of South Africa – part of the Succulent Karoo, a region known for its abundance of unusual succulent plants – are associated with termite mounds, known as 'heuweltjies'. Compared to their surroundings, these heuweltjie soils are deeper, they have more nutrients, and have a better capacity to hold water. The termite nests appear to play an important role in the localized cycling of energy and nutrients. The plant life associated with heuweltjie soils is distinctive, along with a range of small creatures, such as burrowing bees, ants, and mole-rats.

These large termite mounds have been part of the Namaqualand landscape for a very long time. Dating of some heuweltjies has put them in excess of 20,000 years old. Older fossilized mounds extend the links between heuweltjies and Namaqualand ecology considerably further back in time.

Wildlife patterns in the landscape vary over time, of course. The section at the end of Chapter 1 detailed some of the climatic changes that have occurred in many of the world's deserts, changes that naturally also incorporated significant alterations to the plants and animals that existed in these regions. Some of the effects of these changes in climate can still be identified in the flora of certain landscapes today. The plant life of the northeastern Arabian Peninsula (northern Oman and the United Arab Emirates) has distinct similarities with the floras of Iran and southwest Pakistan. This is probably because at some time or times in the past few hundred thousand years, the two regions were connected by a land bridge as global sea levels were lower and the Arabian Gulf region dried up.

Over much shorter timescales, reference has already been made in this chapter to the pulses of activity, driven by the occasional rainstorm, that take place in desert ecosystems. Periods of heavy rainfall produce the spectacular ephemeral desert blooms much loved by nature films and magazines. They also kick-start a multitude of related phenomena and events, including the sudden appearance of ponds and flowing water with their own array of short-lived aquatic organisms, a profusion of insects and seeds, and the emergence of large numbers of other small creatures. Short though these periods of abundance may be, their effects last much longer, as a seed-pulse that drives the entire desert food web for years. It is an appropriate expression of the pre-eminent role played by water as the primary limiting resource in the desert environment.

Chapter 4
Desert peoples

Deserts are home to many people – about a billion in total.
Indeed, the variety of societies living in desert environments is
probably as great as the diversity of deserts themselves. Some of
these groups have long histories of a dryland lifestyle, others are
more recent settlers; all are bound by the common basic needs
of humankind – including water, food, and shelter – but while
traditional communities had to secure these essentials from the
environment around them, many people in deserts today are
entirely dependent on imported resources.

Traditionally, desert inhabitants followed one of three types of
livelihood: hunter-gathering, pastoralism, or farming. Hunter-
gatherer groups such as the Topnaar of the Namib have developed
an extensive knowledge of desert plants and wild animals.
Herders such as those in Mongolia's Gobi desert are more mobile,
making use of domesticated animals – camels and goats – to
produce milk, meat, and leather. Traditionally, desert agriculture
occurs largely around oases and along rivers, though ingenious
means have also been developed to bring water from nearby
mountains via hand-dug underground channels and to harvest
any surface water from rainfall as and when a storm occurs.

Water is the most valuable resource for desert peoples, but other
desert commodities also have a history of use dating back thousands

15. Camel herders in the Gobi desert, Mongolia

of years. These include minerals such as salt and products harvested from desert trees such as frankincense and gum arabic. In recent times, new technologies have enabled society to exploit deposits of minerals deep beneath the desert surface. Crude oil and natural gas are the most significant. Reserves of these fossilized energy sources have been exploited in numerous desert areas, and some states with particularly large reserves – including Kuwait, Saudi Arabia, Qatar, and Libya – have become very wealthy.

The search for oil and gas has led to other discoveries. Sediments that fill the ancient basins beneath some of the shield deserts have been found to contain large amounts of fossil groundwater, particularly below parts of the Sahara and Australia. Modern exploitation of these non-renewable reserves, most built up in the past 100,000 years, has opened up some deserts to novel developments, including large-scale irrigated agriculture.

The spread of agriculture into deserts has also been made possible by technological advances enabling us to transport water across large distances. Many desert cities now benefit from similar pipelines bringing vital water resources from more humid zones. The connections between the inhabitants of these settlements and their desert surroundings have become more and more tenuous, changes that cannot necessarily be sustained in the long term.

The relationships built between various cultures and the desert environment are numerous and diverse. A simple division can be identified between outsiders who see deserts as empty and inhospitable, and those with a better understanding of the desert who know it as home. It is the societies with a proper appreciation of the dryland realm that manage to live in deserts successfully over the long term.

Traditional desert life

The most basic human survival strategy for desert life, and indeed life in all environments, involves living on wildlife that can be hunted or gathered. All early human societies were based on hunting and gathering, obtaining their food from the bounty of nature, gathering wild plants and hunting animals. Such societies were mobile – moving when food supplies in an area dwindled – and relatively small, perhaps consisting of several, often related, family groups.

Success as a hunter-gatherer is totally dependent upon an intimate and detailed knowledge of desert flora and fauna and a familiarity with the sources of water available in a region. This awareness includes experience of how variable rainfall replenishes ephemeral water bodies but also how it affects the growth and behaviour of plants and animals.

Desert hunter-gatherers who continue to practise at least some of the basic tenets of this ancient way of life in the 21st century include a number of Aboriginal groups in Australia, such as the Spinifex people, or Pila Nguru of the Great Victoria desert; the

Topnaar of the Namib; and the !Kung or San bushmen of the Kalahari (the ! stands for a click-like sound that does not appear in English).

Research into hunter-gatherer lifestyles indicates that in many deserts certain types of food resources are particularly important. In southern Africa, the mongongo nut and !nara seeds provide staple foods that can be eaten in season or stored for lengthy periods. The abundant and nutritious mongongo nut is the principal plant food for the !Kung, comprising as much as 50% of their diet. Such a staple is supplemented by a wide variety of other available foodstuffs. The Kalahari provides the !Kung with more than 100 different species of food plant and 55 edible animal species, an inventory that may have been more extensive in the past. In the neighbouring Namib, the Topnaar have food and medicinal uses for more than 80 plant species but !nara seeds – the seeds of the !nara melon – provide the staple diet for a considerable part of the year.

While some plants provide key food resources, others have multiple uses. An example from Australian deserts is the witchetty bush. This shrub is particularly important because of the large tasty grubs – the larvae of a moth – found in its roots, but the seed of the witchetty bush is also a significant foodstuff. The stems and roots are used to make spears, and the leaves and inner bark both have medicinal uses.

For the inhabitants of coastal deserts, seafood may provide the dietary staples. Archaeological finds from the Atacama suggest that the Chinchorro people who settled parts of the desert around 9,000 years ago fashioned fishhooks from bone, shell, and cactus spines, and wove nets using fibres from a tree that grew in dry river beds. Stone spearheads were carved to make harpoons for hunting sea lions, and animal ribs were employed to gather shellfish. Chinchorros set up camps near brackish springs or in lomas fog oases, taking advantage of the plants and associated animals to supplement their marine diet. Some archaeologists think the Chinchorro also collected water where the fog droplets condensed on flat cliff faces.

Groups of desert hunter-gatherers are few and far between in the modern world. Most have become integrated to varying degrees into more contemporary lifestyles. A majority of !Kung living in Namibia work on farms, while others have found employment in wildlife conservation and tourism. Their knowledge of the desert environment is a great advantage in their work and many still use natural desert products in traditional ways, so saving their monetary income for things that are more easily purchased.

The future for the world's few remaining hunter-gatherers is particularly precarious when more powerful groups covet their land for some reason. The potential for diamond mining in the central Kalahari has prompted the government of Botswana to relocate the area's !Kung population, allowing mining companies unimpeded access. Aboriginal rights have been an important issue in Australia in recent times. The Native Title Act of 1993 has protected the legal rights of the Spinifex people to hunt and gather across some 55,000 square kilometres of the Great Victoria desert, although their rights do not extend to minerals and petroleum.

People who herd animals are much more numerous today than hunter-gatherers. Most of the groups that rear livestock inhabit the desert fringe, areas with enough vegetation – mainly grasses – to support domesticated animals but which are generally too dry to sustain cultivation. These areas of grazing are referred to as rangeland.

The lifestyles of these pastoralists are well geared to the changeable environment of desert margins, mobility being an important way of coping with the sporadic and unreliable nature of rainfall and other resources. Mobile pastoralists live in portable structures, usually tents, which can be dismantled easily and moved when necessary, particularly to find new grazing. Like hunter-gatherers, they have few material possessions because everything has to be transportable.

The animals herded are principally sheep, goats, cattle, horses, and camels. The Bedouin tribes of the Arabian Peninsula traditionally herd sheep and camels, plus a few horses. The sheep and camels are kept for their milk, hides, and occasionally meat, the horses for riding and prestige. Pastoralists in eastern and southern Africa keep cattle. Mongolian pastoralists specialize in horses, although camels are more common in drier parts of the Gobi.

The types of livestock have been chosen and bred for their abilities to withstand the rigours of desert-marginal life. Hardy Bedouin goats and fat-tailed sheep are common components of Middle Eastern herds. Other animals may be kept in particular circumstances. For example, many pastoralists in Tibet, and the more elevated parts of Mongolia, herd yaks, which are well adapted to conditions at high altitudes. South American herds at high altitude are made up of llamas and alpacas – the New World camelids.

Their animals provide pastoralists with many items for everyday use. Camels, cattle, sheep, and horses are milked. The milk is drunk and processed into other consumables such as cheese, curds, butter, yogurt, and even alcohol. In East Africa, cattle are bled through a neck artery and the blood cooked or mixed with milk. Although pastoralists eat meat on special occasions, they do so infrequently.

Camels, sheep, and goats produce wool that can be spun to make rope, processed to make felt, or sold at market. An animal's fleece makes a warm rug or coat; its leather is used to make all sorts of items, including bags, ropes, saddles, boots, and shoes. Its hooves can be boiled to make glue. The horn of a goat or cow can be fashioned into a spoon or a handle for a knife.

Animals are also walking fuel dispensers. Burning dry dung makes an excellent fire. This is important in desert margins where trees are often scarce, so fuelwood is difficult to find. Mongolian herders say they use everything but the animal's breath.

Numerous traditional methods have evolved to manage the use of rangelands, to prevent overuse and to ensure some pastures are set aside for grazing during periods of drought. In the Arab world, a system of grazing reserves, the *hema*, has been operational for many centuries. These areas are owned by particular tribes who dictate how they should be exploited. Grazing may be allowed in one *hema* but only at certain times of the year. In another area, grazing may be allowed all year round but the types and numbers of animals permitted are specified for different periods. A limited number of reserves are for beekeeping, but grazing restrictions are relaxed once the flowering season is over.

The *hema* system also had room for some idiosyncratic uses. A unique type of *hema* near Damascus in Syria was maintained for 500 years solely for the use of aged or unhealthy horses, until it was disbanded in 1930. In recent decades, this traditional way of managing pastures has been abandoned in many other regions as social change has led to a breakdown of tribal structures and rangelands have been opened to free grazing by central governments, not always with beneficial results.

Pastoralists maintain a relationship with settled communities, trading animal products for cultivated food and manufactured items. The Drokba nomads of the Changtang, the northern part of the Tibetan Plateau, have always eaten barley flour, *tsamba*, as a staple in their diet. Anthropologists suggest that *tsamba* provides about half of the nomads' calories, but barley simply cannot grow on the Changtang. The Drokba obtain barley, and tea, in exchange for the products of their herds. Yak's wool and tails have long been important items for trade. The tails, set into handles to make fly-whisks, were used in ancient Rome and are still sold in India today.

16. Crops cannot grow in the high-altitude desert of Tibet so Drokba nomads trade the products of their herds for any cultivated foodstuffs they need. The yak is the most important animal on the Tibetan Plateau, providing numerous benefits, including the hair woven to make this tent

Relations between mobile pastoralists and people who live in fixed communities have not always been peaceable. Pastoral nomads have on occasion turned their great mobility to military advantage, raiding towns and villages. The Mongol hordes, the Huns, and the Arabs have historically conquered agricultural civilizations. Such conflicts have continued into recent times. In West Africa, the Tuareg have clashed with central governments of Saharan countries during the 20th century, a period which states in many parts of the world have regarded pastoral nomads as leftovers from a backward era. In consequence, many have attempted to settle pastoralists as a way to assimilate them into a country whose predominantly sedentary lifestyle is regarded as superior to any requiring movement.

Projects to encourage, cajole, and entice nomads to settle, from Sahelian Africa, through the Middle East, to Central Asia and

China, have met with mixed success. In part, this has been due to the fallacy of automatically viewing mobile herding as an outdated way of life. On the contrary, mobile pastoralists are highly resilient people, typically opportunistic and constantly adapting to changing circumstances. They have always had to be resilient in order to endure the environmental vagaries of the desert and its margins.

Certain aspects of a traditional mobile lifestyle have changed significantly for some groups of nomadic peoples. Herders in the Gobi desert in Mongolia pursue a way of life that in many ways has changed little since the times of the greatest of all nomadic leaders, Chinggis Khan, 750 years ago. They herd the same animals, eat the same foods, wear the same clothes, and still live in round felt-covered tents, traditional dwellings known in Mongolian as *gers*. Yet many *gers* now have a set of solar panels on the roof that powers a car battery, allowing an electric light to extend the day inside the tent. Some also have a television set.

In a similar vein, the Bedouin of Syria and northern Arabia still herd sheep, as they have done for many generations, but their use of the camel as a means of transport has been largely replaced by the pick-up truck. The selective modernization of their activities supports a very reasonable standard of living. A study of herding in Syria and Jordan in the 1990s showed that the income during a good year from 300 sheep, a middle-sized herd, was consistently higher than the professional salaries of government civil servants and university lecturers.

The ability of pastoralists to adapt to changing circumstances, which recently have included the collapse of socialism in many countries and the rise of the age of economic globalization, means that they are well placed to continue their lifestyles for generations to come, albeit with some significant modifications. None the less, it is also probably fair to predict that the numbers of people practising traditional pastoralism will continue to decline for a number of reasons. These include attempts by national governments to settle the nomad and dwindling pastures thanks to competition from ranching schemes, increasing numbers of settled farmers, and expanding areas of cultivation.

Cultivation in desert areas also has a very long history. Traditional desert agriculture occurred largely in oases and along rivers where a reliable supply of water is constantly available. Major exotic rivers that cross deserts in northeast Africa, the Middle East, and southwest Asia – the Nile, the Tigris/Euphrates, and the Indus – have been sites for successful irrigation agriculture for several thousand years.

In the valley of the Nile in Egypt, irrigated cropland received water every year from the annual flood in the late summer months. The water that flowed over the riverbanks and across the fields was filled with rich silt, brought from Ethiopia upstream, which helped to maintain the fertility of cropland. At other times, water was raised from the river channel using a variety of clever lifting devices. These included the *shaduf*, a see-saw mechanism with a weight at one end that lifts and swings a skin bag or bucket of water from river to field, and the Archimedes screw, a hollow pipe with a screw mechanism inside.

Elsewhere, other ingenious feats of engineering were developed to support irrigated agriculture using groundwater. One of the most significant of these probably first originated in Persia getting on for 3,000 years ago. It is an underground system of gently sloping channels that convey water from an aquifer to lower-elevation farmland. The flow of water through the tunnels, all dug by hand, occurs entirely thanks to the force of gravity, so there is no need for pumps. Being almost completely underground, the amount of water lost to evaporation is minimal. Some of these underground conduits carry water for tens of kilometres. In Persia, and in modern Iran, this type of underground channel is known as a *qanat*. The same technology is found in desert areas all across Asia, in North Africa, and in Spain. They are known as *falaj* in Oman, *fogarra* in North Africa, and *karez* in Pakistan, Afghanistan, and northwestern China. Some of these systems, which need periodic maintenance via vertical shafts sunk at intervals of 20 to 30 metres, are still in use after 1,000 years and more. The oasis of Turpan in northwestern China has long been famous for its grapes, grown with the aid of around 1,000 *karez* that total about 5,000 kilometres in length. The system has been channelling underground water to the Turpan depression from nearby alluvial fans for at least 2,000 years.

Several techniques have also been developed by resourceful desert inhabitants to harvest what little surface water is available in dryland regions. Many are based on collecting, guiding, and ponding storm water, often by the use of earthen or stone walls. The effect is to concentrate water into relatively small areas which are cultivated or used to graze livestock. Many of these ancient techniques are well documented from deserts in South America, the arid southwest of North America, and across North Africa and the Middle East. At its simplest level, this sort of runoff farming or water harvesting involves constructing check dams across normally dry valleys. The dam impedes the flow of water generated by the occasional rainstorm, causing the water to deposit its sediment (helping to build up a soil) and encouraging the ponded water to soak into this alluvial fill. A small check dam might support a single fruit tree, for example. Larger dams may allow a cereal crop to be planted in a small field.

This type of small check dam and its associated terraced area is called a *'jessr'* (plural: *jessour*) in Tunisia. *Jessour* cover an area of about 4,000 square kilometres in the south of the country, allowing crops to be grown in an environment that is otherwise too arid for agriculture. The most common crops are olive trees and drought-resistant annual grains with a short growing period such as wheat and barley.

More sophisticated systems involve long walls that snake across low-angle slopes to channel any storm-generated overland flow of water towards the desired portion of valley floor. A greater flow of water can be generated over a large slope area by the clearance of obstacles to overland flow such as pebbles and boulders. Removing larger obstacles means that soil pores quickly become clogged with fine material when water flows across them, effectively sealing the surface, so reducing the amount of water that infiltrates into a slope and thus generating greater runoff.

Some of these techniques have been used for a very long time. Well-developed runoff farming systems studied at Jawa in Jordan seem to be at least 5,000 years old, and those at a site called Beidha in the Edom mountains in southern Jordan may have been in use nearly 9,000 years ago. In North America, the Papago Indians have lived for centuries in the Sonoran desert, in the US state of Arizona and the Mexican state of Sonora, practising a desert agriculture based on water harvesting. Traditional farmers in the semi-arid Zuni area of the US southwest (New Mexico) have successfully cultivated maize and other crops in this way for more than 3,000 years. The capture and redirection onto fields of organic-rich sediment by Zuni farmers also means that they have no need to apply artificial fertilizers in the contemporary era.

A similar range of techniques has also been employed by traditionally nomadic people in the Karakum desert in Turkmenistan to secure water supplies for consumption by their livestock and to produce some crops for themselves and their animals. These methods focus on flat or slightly sloping clay surfaces called '*takyrs*'. *Takyrs*, which have little or no vegetation, act as natural catchment areas. Several types of hand-dug basin that collect rainwater runoff have a long history of providing temporary storage points for watering livestock on *takyr* surfaces. In some cases, dome-shaped brick covers have been built over these small reservoirs to reduce evaporation. Another technique is to increase the productivity of small depressions in the *takyr*, where water collects naturally, by digging trenches to channel a greater flow of runoff. The depressions, or '*oytaks*', have a sandy topsoil and are used for hay production, but crops such as melons and gourds are also grown. Typically, a cultivated *oytak* of 10 square metres requires at least 1,000 square metres of catchment to provide it with adequate moisture.

Contemporary desert life

Societies in all parts of the world are dynamic, adapting to changing circumstances both in the natural environment and in the ways society itself is organized and the ideas and technologies it has to work with. Some groups continue in their traditional ways while making certain adaptations to developments taking place around them. The degree of adaptation varies considerably from place to place.

In numerous countries with desert regions, the need to feed growing human populations has spawned large-scale projects to expand and intensify food production in areas perceived by many in authority to be empty and worthless. Deserts can certainly bloom given enough water, so engineers in many parts of the world have developed methods of transporting water to desert regions from areas with ample supplies. Several projects to bring water to the Thar desert in India from rivers to the east have been implemented over a period of nearly 100 years since the launching of the Gang canal in 1927. The huge Indira Gandhi Canal scheme comprises nearly 8,000 kilometres of distribution channels which have allowed large areas of gently undulating arid grasslands to be transformed into agricultural landscapes.

Similar huge expansions in irrigated cropland were achieved in parts of Central Asia during the 20th century during the period of the Soviet Union. The Karakum Canal, which diverts water from the Amudarya westwards across southern Turkmenistan, was completed in 1986 after more than 30 years in construction.

The canal is no less than 1,400 kilometres long and the water it delivers has enabled crops to be grown on more than 7,000 square kilometres of land that would otherwise be too dry for cultivation. Sadly, however, Turkmenistan's irrigation systems are poorly designed and inefficient, resulting in large tracts of cropland facing problems of salinization and/or waterlogging. The loss of water from irrigation canals is also considerable. Most are not lined, so water seeps away, and evaporation rates in the arid climate are high. Consequently, more than one-third of the water diverted from the Amudarya never reaches the fields. The disastrous offsite impacts of Central Asia's vast irrigated area on the Aral Sea are outlined below (see Desertification).

Transfers of water on this scale have been made possible by technological advances in several fields, not least in earth-moving and concrete technology that have allowed construction of larger and larger storage dams. People have constructed dams to manage water resources in arid regions for thousands of years. One of the oldest was the Marib dam in Yemen, built about 3,000 years ago. Although this structure was designed to control water from flash floods, rather than for storage, the diverted flow was used to irrigate cropland.

The modern era of large dams, many more than 150 metres in height, dates from the early 20th century. These enormous structures have increasingly been designed for multiple purposes, including agricultural irrigation, water supply for industrial and domestic use, hydropower generation, and flood control. The environmental changes caused by large dams have been significant and not always welcome.

The Aswan High Dam, completed in 1970 on the Nile, generates about 20% of Egypt's electricity, protects the country's population from both floods and droughts, and has provided water to nearly 5,000 square kilometres of new cropland, a particularly important aspect for a largely hyper-arid country where just 3% of the national territory is suitable for cultivation. Construction of the dam has required some changes to traditional Egyptian farming methods, however, because it has meant the end to the river's annual flood and deposition of nutrient-rich silts. Farmers now have to use expensive synthetic fertilizers instead. Other problematic downstream effects of the Aswan High Dam have also been linked to the lack of silt arriving at the Nile delta, which has meant more coastal erosion, greater salinization due to the intrusion of seawater, and a decline in the eastern Mediterranean sardine catch.

Although groundwater has been exploited for desert farmland using hand-dug underground channels for a very long time, the discovery of reserves of groundwater much deeper below some deserts has led to agricultural use on much larger scales in recent times. These deep groundwater reserves tend to be non-renewable, having built up during previous climatic periods of greater rainfall. Use of this fossil water has in many areas resulted in its rapid depletion.

Extensive spraying systems involving a lengthy boom that rotates on wheels around a central pivot have been developed to use this groundwater. The resulting marked circles of vegetation, up to a kilometre in diameter, irrigated by these centre-pivot systems have become a common sight in many desert and semi-desert areas. One such area is the High Plains of the US southwest, where water for irrigation from the Ogallala aquifer has helped to create one of the largest areas of intensive agricultural production on Earth. Irrigation using the Ogallala aquifer has a limited lifespan, however. For many years, water has been pumped from the aquifer at rates that have far exceeded the rate of recharge. Although some centre-pivot systems use low-energy, precision-drip application and even subsurface drip technologies designed to save water, the number of wells has continued to rise, further depleting the aquifer.

Some very large agricultural schemes have appeared in southern Libya in recent decades, using fossil water from beneath the Sahara in the Murzuq and Kufrah basins. The Irawan agricultural project in the southwest of the country covers 500 square kilometres in one of the most hyper-arid parts of the Sahara, meaning that natural recharge of the Murzuq aquifer is considered to be non-existent. However, the greatest challenges faced at Irawan have been social. In consequence, Libya has implemented its so-called Great Manmade River project, a vast scheme to transport water from sub-Saharan groundwater reserves across thousands of kilometres of desert to the coast, where most of the population prefer to live. No one is sure how long this enormous engineering scheme, financed by the country's great oil wealth, will continue before the groundwater either runs out or becomes too salty to use – some estimates suggest only a few decades. As in the US High Plains, Libya's groundwater schemes do not have a long-term future.

Early settlers of the semi-arid High Plains did not use irrigation but simply ploughed up large areas of grassland to create farms (where cycles of drought and crop failures eventually culminated in the devastating Dust Bowl of the 1930s – see below). The conversion of semi-desert areas to cropland without the aid of irrigation has also occurred in many other parts of the world. Very large areas of grassland have been replaced by fields of cereals in the prairie provinces of western Canada, the steppes of Kazakhstan, and the pampas of Argentina. Many areas of converted desert fringe grassland also support ranching, a more intensive approach to rearing livestock than traditional mobile pastoralism. Other dryland ecosystems have also been threatened by the steady spread of agricultural land. In the semi-arid southwest of Madagascar, much of the unique spiny forest has been lost to a combination of slash and burn farming and plantations of sisal.

The production of food is just one of many activities people undertake in desert areas today. In some parts of the world, new lifestyles and societies have arisen, made possible by novel ideas and technologies. Pre-eminent among these developments is the ease with which we can move people and resources, often over great distances and with considerable speed. Increasing numbers of people simply visit desert areas rather than live there. Many forms of tourism and recreation have flourished in deserts, with focuses as diverse as adventure experiences, ecotourism, sporting, cultural, and seaside holidays. The countries of North Africa, particularly Morocco, Tunisia, and Egypt, have developed all of these forms of tourism to become important earners of foreign exchange, welcoming predominantly European visitors.

Other groups of outsiders with no traditional ties to deserts have also recognized the attractions of a dry climate, almost continuous sunshine, wide-open spaces, and magnificent scenery. Recent decades have seen substantial population growth in the drylands of the US southwest, part of the so-called Sun Belt, fuelled in large part by people arriving from other parts of North America. Living in the deserts and semi-deserts of the USA has been made more bearable for these people by the advent of air conditioning. Some older, more affluent Americans live in desert communities entirely dedicated to those who have retired from paid employment.

By contrast, very large, remote, and sparsely populated areas of deserts and their margins have been used for modern-day purposes that would be difficult or inconceivable in almost any other region. The first nuclear bomb was tested in the Chihuahuan desert in 1945, at La Jornada del Muerto in the US state of New Mexico. The British tested their nuclear weapons at Woomera in South Australia and the French at Reggane in the Sahara. China used Lop Nor and the Soviet Union experimented at Semipalatinsk, in the northeast of Kazakhstan today. Sites in Baluchistan in western Pakistan and at Pokharan in India's part of the Thar desert have been used for similar purposes. Many of these extensive desert areas remain in the hands of the military. La Jornada del Muerto is today part of the White Sands Missile Range, the largest military installation in the USA. The Woomera

Prohibited Area, administered by Australia's Department of Defence, covers 127,000 square kilometres of desert, an area almost the size of England, making it the world's largest land aerospace testing range.

Desert cities

Sizable cities have been found on desert margins for millennia. These settlements developed on trade routes or at sites of religious significance. Some have a long history of continuous occupancy and importance, though others have not adjusted successfully to changing circumstances. Baghdad and Cairo ranked among the world's largest cities at both the beginning and end of the second millennium. Other cities of major importance a thousand years ago, such as Nishapur in Persia, have now faded from prominence. Much newer cities, including Dubai and Las Vegas, have emerged to take their place as regionally important urban centres in the desert.

Most authorities agree that the first urban cultures began to develop about 5,000 years ago in Egypt, Mesopotamia, and India, and there is some support for the theory that the creation of the city and the process of urbanization was at least in part a product of the need to organize irrigation in desert areas, a notion explored further in Chapter 5. Both Damascus and Sana'a regularly claim to be the world's oldest continuously inhabited city. Jericho is the earliest known walled city, but the settlement has probably been abandoned several times through its history.

The first desert cities grew slowly, in relative harmony with their dryland surroundings. They could only develop in places with a local, accessible water supply. Most buildings were constructed of clay or mud bricks, with thick walls to restrict the conduction of heat into the interior. Large-scale urbanization in desert areas is a much more recent phenomenon, one coupled with very rapid growth in some of today's desert cities.

Certain deserts, in Arabia and North America particularly, have seen rapid urbanization linked to the economic wealth generated by mineral exploration, which has paid for improved water-management techniques to provide sufficient water for modern cities in the desert. Many cities in Sahelian Africa have experienced similarly rapid growth, largely due to the influx of migrants whose rural livelihoods have been rendered impracticable by drought.

Oil has, more than any other natural resource, been the driver of urban growth over the last 50 years or so, initially in North America and later in Arabia and Saharan Africa. The prosperity generated by desert oil fields fuelled rapid urban development in the oil-producing states of Saudi Arabia, Kuwait, and Libya. Those countries, including Egypt and Jordan, that provided labour, political support, and transit facilities for oil also experienced a brisk pace of urban expansion.

Cities in several Arabian Gulf states have seen dramatic changes in their economic, social, and cultural character thanks to oil wealth. Dubai has been transformed from a fishing settlement of little significance to a cosmopolitan 21st-century city of regional importance in just five decades; the city has made the transition from pre-industrial, through industrial, to post-industrial status in only 50 years, a process that in many Western cities took two centuries or more. The challenges of providing adequate infrastructure during this period of rapid urban growth have been eased by the large revenues earned by oil exports. Dubai, like other Gulf state cities, also has the advantages of ample land resources in an empty desert hinterland, no legacy of industrial dereliction, and no sprawl of spontaneous settlements to hamper urban development. A reliable supply of water, a perennial issue for any city but particularly for those situated in deserts, has also been resolved with an expensive solution. Most of Dubai's water is produced by the desalination of seawater, a very energy-intensive operation.

17. Recent rapid, large-scale urban development in Dubai has been financed by revenues derived from oil exports

Development of modern health-care facilities have played their part in boosting the population of Dubai, pushing infant mortality rates down and overall life expectancy up, but a vast wave of immigration, an influx of foreign labour and expertise, has also been essential to Dubai's urban trajectory. In 2005, more than half of the city's population was born abroad.

Immigration on a large scale, but of a rather different kind, has also been important in the recent expansion of Mauritania's capital, Nouakchott. In 1955, five years before the West African country's independence, Nouakchott was a fortified village with about 1,800 inhabitants. Mauritania's new capital had become a city of more than 40,000 people by 1970. Over the next three decades, a period when prolonged drought wracked all the countries of the Sahel, Nouakchott experienced a veritable population explosion. Although estimates vary widely, because most of the population growth was completely unplanned and unregulated, the city's population had passed the 500,000-mark by 2000. This staggering rate of growth has continued. By 2005, Nouakchott's population was probably 750,000, but some suggest it was closer to 1 million.

Water supply has long been a critical issue for Nouakchott. The influx of environmental refugees, escaping the prolonged drought in the Sahel, was much too fast for the city's authorities to keep up and provide Nouakchott's new districts with water tanks and public standpipes. For more than 15 years, the city's hydraulic installations have been providing water for ten times the number of people they were designed to serve. Nouakchott's water is supplied from an aquifer via a 60-kilometre pipeline. Some of the capital's working-class neighbourhoods face rationing and daily cuts in supply, but a large proportion of the city's population is concentrated on the outskirts of the city where there is no municipal supply of either water or electricity.

The consumption of water in Nouakchott is very low, largely because of the poor supply. On average, each person consumed 40 litres a day in 2000. Dubai, by contrast, has one of the highest per capita rates of water consumption in the world. On average, each inhabitant of Dubai consumes 50 times more water than his or her counterpart in Nouakchott.

A new scheme to provide water to Nouakchott from the Senegal River 200 kilometres away is in progress. Expanding cities in many deserts have been compelled to look further and further afield for their water, often causing significant impacts in the source areas. Tehran is a case in point. Before the 1930s, the city's water was provided by a series of traditional *qanats*, but since that time the population of Tehran has grown from some 300,000 to more than 8 million people.

The acceleration in Tehran's need for water spurred the construction of a series of dams and canals in the early decades of the last century, to bring water 50 kilometres from the Karaj

River to the west, but these schemes reduced the water available for rural agriculture in the process. By the 1970s, supplies were again running low, so water was diverted more than 75 kilometres from the Lar River to the northeast. Rapid economic growth during the 1970s and 1980s, and a pulse of migration into the city after the Islamic revolution in 1978, spurred exploitation of local groundwater supplies to meet the continuing rise in demand, but the quality of this resource is under threat from the city's inadequate system of waste-water disposal, which is sent underground, without any treatment, through the use of injection wells to recharge the groundwater.

In the southwestern USA, demand from dryland cities and agriculture has virtually sucked dry the Colorado River, the region's major waterway, which was dammed in several places during the 20th century. Almost all of the Colorado's water is exported to other areas for use in growing cities and on farms. In the state of Colorado alone, 10 tunnels have been drilled through the Rocky Mountains to pipe the water eastward to population centres. Similarly, a huge aqueduct takes Colorado River water nearly 500 kilometres across the Mojave desert to farming areas and large cities on the west coast of California, including Los Angeles and San Diego. The numerous environmental impacts include the loss of many of the river's native species of birds and fish. The thirsty city of Los Angeles is also responsible for the complete loss of Owens Lake, 350 kilometres to the north in California, which had disappeared by 1930, leaving 220 square kilometres of desiccated fine-grained lake sediments to become a source of frequent regional dust storms. Only legal action has saved Mono Lake from a similar fate.

Competition for water, between cities and farms and between individual settlements, is intense in the dry US southwest. Other rapidly growing urban areas in the region include Las Vegas, Phoenix, and Tucson. The inhabitants of these desert cities enjoy an oasis lifestyle supported by sophisticated storage dams, water-pumping and distribution technologies. Like many cities in the Gulf states, these settlements maintain an astonishing level of greenness given the aridity of their locations. Much of the urban and suburban landscapes in these cities are more typical of a temperate climate rather than a desert. Verdant lawns, lush plant life, fountains, and swimming pools are maintained in an environment more suited to cacti, succulents, and dry brush. The growth of this type of city, with a mentality at variance with its desert surrounds, is expensive in both economic and environmental terms. It has put unsustainable pressure on natural resources and cannot continue in its present form for very much longer.

Deserts and culture

The long history of human association with deserts has had a wide variety of cultural consequences for many societies. These inevitably include cultures that have developed inside drylands, but also many that have emerged far from the desert realm.

The Aboriginal culture in Australia is perhaps more intimately entwined with the desert landscape than most. The identity of all Aboriginal peoples is linked to a particular geographical site which is their spiritual home. The landscape itself was created by ancestral beings who, once their work was done, seeped back into the earth or transformed themselves into topographical features such as rocks, trees, or waterholes. These features continue to be revered as hubs of supernatural power.

Similarly, other landmarks are associated with an exploit or episode in a sacred myth, or with a verse in a sacred song, so that for Aborigines the entire desert countryside is imprinted with a spiritual meaning that is invisible to outsiders. To give just one example, the Walmajarri people of the Great Sandy desert believe that the bare plains lying between the parallel linear dunes were formed by two giant mythical snakes that wiped the landscape clear while travelling westward through the dunes. (These areas are explained in a scientific culture by invoking the action of the wind.)

Connections between the topography of a landscape and the ancestral spirits that created it are continually reinforced by the singing of Aboriginal songs that trace the narratives of journeys undertaken by the ancestors. The more mundane, practical value of these songs is as verbal maps to indicate paths for tribal groups to follow, to find food and water or to perform ceremonial rites.

The Aboriginal view of the desert, as a place full of spiritual richness and life, stands in stark contrast to the perception prevalent in many other cultures of the desert as an empty place of alienation and punishment. This perspective is embodied in the Christian Bible's story of the Holy Spirit leading Jesus into the desert, where he went without food for 40 days and nights before the devil arrived to test him.

Similar depiction of the desert, effectively as a metaphor for Hell, occurs in the Koran, the holy book of Islam. Heaven is frequently described in the Koran as a lush garden with babbling brooks, whilst Hell is a place of blazing heat, where the deeds of sinners are like a mirage: briefly exciting but ultimately empty.

The dangers lurking in deserts, embodied as ogres, demons, and djinns, also feature prominently in Persian legends and poetic epics. One group of ogres that preys on travellers does so by jumping on to a man's shoulders and throttling him with its legs. The 12th-century Persian philosopher Suhravardi explains the only way to deal with this hazard is to board Noah's Ark and grab hold of the staff of Moses.

At night, however, the Persian ogres and demons are replaced by benevolent fairies and the desert is also celebrated in many cultures for its serenity and beauty. Much of the poetry that comprises an integral part of classical Arabic literature is focused on the desert and its natural wonders, with passionate descriptions of desert plants, animals, and dunes, the burning sun and shivering cold, the moon and the stars. Virtually all pre-Islamic long poems contain a section on the desert. The Koran also has room to praise the arid environment as a beautiful part of Allah's creation.

The magnificence of the desert realm, its exotic and otherworldly nature, has featured prominently in the literature and art of Western nations over a period that extends back many centuries. The European perception can be traced at least as far as Marco Polo's tales of derring-do traversing the Gobi and Taklimakan that became very popular in parts of Europe during the 13th century. In more recent times, novelists, poets, artists, musicians, and film-makers from Europe and North America have been inspired by the world's deserts as places of mystery, silence, and contemplation, as well as isolation, suffering, and hardship.

Western writers over the last century whose work is associated with deserts include T. E. Lawrence, Antoine de Saint-Exupéry, Paul Bowles, and Edward Abbey. Deserts have also featured prominently as vast, harsh, and hostile places in several branches of world cinema. Film-makers who have shot some of their most notable works in desert landscapes include the Americans John Ford and Sam Peckinpah, the Italian Sergio Leone, the Briton David Lean, and the Australian Peter Weir. Their views of the desert as a setting for epic human struggles have helped to shape contemporary cultural perceptions of drylands in many Western minds. Consequently, in the USA and to a lesser extent in Australia, the desert, in combination with attempts to survive against the odds in the arid wilderness, has become symbolic of the national character.

Desertification

Numerous societies have occupied deserts and their margins for thousands of years. Over this time, people have developed ways of living that are attuned to the vagaries of desert environments. Obviously, human activities have also had great effects on desert landscapes. Part of the shifting relationships people have had with drylands over the centuries has included instances when productive land has been lost to the desert. In some cases, the cause has been human, through overuse and mismanagement of dryland resources, while in other cases natural changes in the environment have reduced the suitability of such areas for human occupancy. Often a combination of human and natural factors is at work.

The term 'desertification' has been coined for the loss or degradation of natural resources in drylands. This phenomenon is by no means a new one, but desertification was arguably the first major environmental issue to be recognized as occurring on a global scale. That recognition came in the early 1970s. Today, desertification still threatens the usefulness of desert ecosystems in many parts of the world, although the issue has also become imbued with controversy, including disagreements over the actual magnitude of the problem worldwide.

Significant human impacts are thought to have a very long history in some deserts. One possible explanation for the paucity of rainfall in the interior of Australia is that early humans severely modified the landscape through their use of fire. Aboriginal people have used fire extensively in Central Australia for more than 20,000 years, particularly as an aid to hunting, but also for many other purposes, from clearing passages to producing smoke signals and promoting the growth of preferred plants. The theory suggests that regular burning converted the semi-arid zone's mosaic of trees, shrubs, and grassland into the desert scrub seen today. This gradual change in the vegetation could have resulted in less moisture from plants reaching the atmosphere and hence the long-term desertification of the continent.

Some of the most clear-cut examples of desertification are those that have occurred on farmland because the resulting declines in crop yield are relatively straightforward to monitor. Fields on which just a single crop is grown year after year, so-called 'monocultures', will slowly become degraded, as studies on cropland in the semi-arid Pampas of Argentina have shown. The long-term cultivation of millet has affected both the chemical and physical properties of soils. The depletion of nutrients means that larger amounts of fertilizers have to be applied to maintain crop yields, while declines in organic matter and soil stability have meant a greater susceptibility to erosion.

Erosion of soils on desert-marginal fields is a classic manifestation of desertification. In many cases, erosion has been caused by the introduction of mechanized agriculture, with large fields and deep ploughing. These areas may produce decent returns for farmers during years of good rainfall but they can be prone to severe soil erosion during times of drought.

The Dust Bowl disaster of the 1930s on the US High Plains is one of the most notorious examples. Spectacular and devastating wind erosion occurred on a huge scale, the winds carrying soil dust northwards as far as Canada and eastwards to New York and out over the Atlantic Ocean. At Amarillo, Texas, at the height of the Dust Bowl period, one month had 23 days with at least 10 hours of airborne dust, and one in five storms had zero visibility. The large-scale environmental degradation, combined with the effects of the Great Depression, ruined the livelihoods of hundreds of thousands of American families.

The build-up of salts, which are harmful to plants and the texture of soils, is another common facet of the desertification problem. It can be particularly acute on poorly managed irrigation schemes, and one estimate suggests that nearly half of all the irrigated land in arid and semi-arid regions is affected to some extent by this issue. In its most advanced stages, salinization means that no crops can be grown at all, rendering fields totally useless to the farmer.

In Pakistan, where irrigated land supplies more than 90% of agricultural production, about one-quarter of the irrigated area suffers from salinization. The issue is particularly acute on the cropland irrigated from the Indus, the largest single irrigation system in the world. The costs of salinization in Pakistan are undoubtedly high; the problem depletes the country's potential production of cotton and rice by about 25%.

Salinization is just one of a suite of difficulties that have dogged irrigated agriculture in Central Asia and combined to produce an environmental disaster in and around the Aral Sea since the 1960s. Diversion of water from the region's two main rivers, the Amudarya and Syrdarya, has meant a dramatic decline in the amount of water flowing into the Aral Sea, while evaporation has continued at its normally high level thanks to the region's desert climate. The Aral Sea has shrunk in consequence. In 1960, it was the fourth-largest lake in the world, but since that time its surface area has more than halved, it has lost two-thirds of its volume, and its water level has dropped by more than 20 metres. In some parts, the Aral Sea's remaining waters are more than twice as salty as seawater in the open ocean.

These spectacular changes have had far-reaching effects. Most of the Aral Sea's native organisms have disappeared, unable to survive in the salt water, meaning an end to a once-major commercial fishing industry. The delta areas of the Amudarya and Syrdarya have been transformed due to the lack of water, affecting flora, fauna, and soils, while the diversion of river water has also resulted in the widespread lowering of groundwater levels.

Receding sea levels have had local effects on climate, and the exposed seabed has become a source of dry sediments that are blown over surrounding agricultural land up to several hundred kilometres from the sea coast. This dust is laden with salts, adding to the problems of irrigated agriculture. It is also thought to have adverse effects on human health.

18. Abandoned trawlers near the Aral Sea in Kazakhstan, victims of perhaps the most dramatic example of human-induced environmental degradation in the modern era

The other major land use in deserts and semi-deserts, pastoralism, has also been implicated in many cases of desertification. Grazing animals may cause degradation in several ways. Increased soil erosion, by wind or water, may result from a loss of vegetation cover caused by heavy grazing pressure in combination with other effects of livestock such as trampling and the compaction of soils.

Another widespread effect of intensive grazing is the encroachment of unpalatable or noxious shrubs into rangelands. High cattle densities are thought to encourage invasion of grasslands by thorny bushes in the Kalahari. Cows tend to avoid these bushes because of their thorns and these species become more abundant as a result. A thick cover of thorny bushes deters the growth of grass and this type of bush encroachment has resulted in a significant reduction in the extent of high-quality rangeland in Botswana.

In some parts of the world, degradation has been the result of a change in the approach to pasture use: from a flexible strategy typical of traditional pastoralists in which the natural dynamism of dryland vegetation is mimicked by regular movement of herds and maintaining several different animal species, to a less flexible Westernized approach in which rangeland may be fenced off and only cattle are grazed. This is the situation in parts of Botswana. In other cases, pointing the finger of blame at herders may be too convenient and more a reflection of the position of traditional pastoralists at the margins of settled society than anything else. Many authorities now believe that rainfall variability is a more important determinant of the health of rangeland and its soils than 'overgrazing'.

One further complicating aspect in the desertification debate is the fact that our ideas of how valuable a desert might be can change. Resources are, by definition, cultural appraisals of things that might be useful to us. A hundred years ago, when large areas of spiny forest in semi-arid Madagascar were cleared to make way for sisal plantations, most people were likely to have considered this clearance beneficial in economic terms. (Small segments of sacred forest were left untouched among the plantations.) Viewed with hindsight, and a set of values that place more emphasis on the wider good of unique ecosystems for their own sake, this action is less defensible.

The biological diversity of many desert and semi-desert areas continues to be threatened by many different human activities. These include farming and herding, as outlined above, but also industrial and urban expansion, fuelwood collection, mining, hunting, and numerous forms of pollution. The planet's biologically richest but also most endangered regions have been highlighted by conservationists as global hotspots. Two of these are wholly within deserts: the Succulent Karoo of South Africa and Namibia; and the Horn of Africa, which includes coastal parts of Yemen and Oman.

19. Agave plants, native to North American deserts, have been introduced to many other parts of the world to be grown in commercial plantations for their strong fibre: sisal. This one in southwestern Madagascar required large-scale clearance of the region's unique semi-arid spiny forest

Southern Africa's Succulent Karoo is renowned for its large numbers of unique plants – particularly succulents – and reptiles that are found nowhere else on Earth. The main threats in this area come from grazing, agriculture, and mining, especially for diamonds and heavy metals. Livestock also present a significant threat to biodiversity in the Horn of Africa, along with the charcoal industry and a general lack of authority from government, particularly in Somalia. Trees native to this area have been producing gum-resins such as frankincense and myrrh for thousands of years. The Horn of Africa is also home to a number of endemic and threatened antelope, notably the beira, the dibatag, and Speke's gazelle.

Elsewhere, some of the world's largest nature reserves have been established in desert areas with very small human populations. The Rub al Khali Wildlife Management Area in Saudi Arabia encompasses a large part of the world's greatest sand desert, and the Great Gobi Strictly Protected Area covers a swathe of predominantly stony desert. Both of these reserves have been the focus of attempts to reintroduce large species that had become extinct in the wild, the Arabian oryx in Saudi Arabia and the takhi or Przewalski horse in Mongolia.

The next chapter goes on to explore some of the broader links between people and deserts. These include ideas and innovations that have been adopted by societies all over the world, as well as connections between the physical geography of deserts and the character of the planet as a whole.

Chapter 5
Desert connections

Deserts do not exist in isolation; they are identifiable as distinct from other parts of our planet, but at the same time they do, of course, remain an integral part of it. Interactions between deserts and the rest of the world, in both the physical and human sense, are innumerable. Recognizing and understanding these desert connections helps us both to appreciate the importance of deserts in the widest perspective and also to gain a more complete knowledge of human society and its natural habitat.

The Earth system

The Earth's climatic zones – arid, temperate, polar, and so on – are simply artificial academic distinctions that we have developed to help us divide and understand the very complicated planet we live on. All these zones interact within the single entity that many physical scientists like to call the 'Earth system'. Any major event in the Earth system sets in motion a whole series of processes that cut across climatic zones and have impacts, both direct and indirect, on desert landscapes.

A change in the Earth's climate is an example. Over the last two million years, a series of glacial and interglacial phases have occurred on Earth. The increases in glacial ice have largely taken place in high latitudes (plus some expansion in mountainous areas nearer the Equator), but marked effects have also been detected in deserts. These effects stem predominantly from the greater aridity that prevailed during these periods.

The legacies of these changes remain in contemporary deserts. For example, lower sea levels during glacial periods resulted in the Arabian Gulf becoming dry. This land bridge affected the distribution of plants we see today. As outlined in Chapter 3, there are distinct floral similarities between the Makran coast of Iran and Pakistan to the north and northern Oman and the United Arab Emirates to the south. Quartz sediments blown from the dry bed of the Arabian Gulf also provided much of the sand that comprises the Rub al Khali in Saudi Arabia.

The effects of drier deserts are also detectable on their contemporary margins. Chapter 1 highlighted the presence of many once-active sand dunes in areas too humid for sand-dune formation today. Enhanced erosion by wind also produced greater quantities of desert dust that was deposited sometimes great distances away. Deposits of this wind-blown silty material, known as 'loess', blanket the terrain on the margins of some deserts. The thickest and most extensive occur in northern China. On the Loess Plateau near Lanzhou, these deposits reach depths of several hundred metres. The soils formed on loess are themselves rather easily eroded, but they are also renowned for their great fertility.

Dust is still blown from the deserts of China and Mongolia today in broadly similar directions: towards the east and southeast. Most of this dust settles out of the air and is deposited close to its source, so the Loess Plateau continues to receive desert dust, although the amounts are not as great as during glacial periods. The longer the distance from the desert source, the smaller the amounts of dust transported, but Asian desert dust is frequently blown as far as Japan, where the dust haze is known as *kosa* (literally, 'yellow sand'). Clouds of dust raised in very large storms in northeast Asia have been traced on satellite imagery, taking about a week to move across the Pacific Ocean to Canada and the USA. Convincing evidence of dust movement over even greater distances has also been found. One study detected small amounts of dust from Chinese deserts in the French Alps, having been transported more than 20,000 kilometres across the North Pacific, North America, and the North Atlantic.

Desert dust undoubtedly constitutes one of these significant connections between deserts and the rest of the planet. It is important for many reasons. Desert dust has an impact on the workings of both terrestrial and marine ecosystems through its effects on soil characteristics, oceanic productivity, and the chemistry of the atmosphere. Largest among all of the world's sources of desert dust is the Sahara, which may produce as much as a billion tonnes of dust every year. The effects of this material are numerous and far-reaching. Saharan dust makes an important contribution to the formation of soils in southern Europe; it carries disease-spreading spores which have been linked to the demise of coral reefs in the Caribbean; large amounts of dust appear to suppress the formation of hurricanes over the Atlantic, and Saharan dust provides critical nutrients to the rainforests of the Amazon Basin in South America.

Biological events in the desert also have external impacts. A dramatic example is provided by periodic plagues of desert locusts. Although normally restricted to semi-arid and arid parts of Sahelian Africa, the Middle East, and southwest Asia, the desert locust can on occasion swarm over much larger areas. When environmental conditions are favourable – a few successive years of relatively warm and humid conditions – locust numbers can grow very quickly, leading to extremely high population densities. In outbreak years, locusts can affect an area of 30 million square kilometres, more than 20% of the land area of our planet.

The impact on all forms of vegetation, including crops, is often disastrous. During plagues, the desert locust has the potential to damage the livelihood of one-tenth of the world's human population. The largest single swarm of desert locusts that was precisely measured was recorded in Kenya in 1954. It covered 200 square kilometres with an average of 50 locusts per square metre. That made a total of 10 billion individuals. The average locust consumes roughly its own weight in fresh food per day, so this swarm ate as much as 500,000 camels, or five million people.

Lastly in this section, some possible wide effects of climate and its changeability are examined. Some of the ways in which mismanagement by people can cause the processes of desertification were outlined in Chapter 4, but it should be remembered that variations and changes in climate can also result in broadly similar effects. Climate changes can modify many aspects of deserts that could be detrimental to human societies. The most obvious dangers, given the importance of water, would be any trend towards greater aridity.

A serious connection has even been made between the long history of aridity in the Atacama desert and the tectonic processes that have created the Andes. The proponents of this theory argue that the link lies in the fact that the hyper-arid Atacama is a desert with very low rates of erosion. Hence, only very small amounts of sediment from the Atacama reach the seabed offshore. This is the Peru-Chile trench, site of the greatest forces between the Nazca and South American tectonic plates which have thrust the Andes upward.

The argument goes that the tectonic forces in this area are especially potent because the boundary between the tectonic plates is not dampened and lubricated by sediment from the adjacent land. In other words, this theory proposes an intriguing reversal of part of the argument put forward in Chapter 1 to explain the Atacama's very dry climate. Rather than uplift of the Andes being one of the primary causes of aridity in the Atacama, by forming a mountain barrier to moist air from the east creating a rain-shadow effect, aridity instead may have promoted uplift.

If confirmed, this explanation would make a very dry desert climate the reason for, rather than the consequence of, mountain building (even though, once begun, positive feedback between increasing elevation of the Andes and a growing rain-shadow effect would have helped to maintain or intensify hyper-arid conditions in the Atacama).

Desert resources

Deserts, their landscapes and resources, have without doubt had a profound influence on the evolution of human societies all across the globe. Natural resources have been exported from certain desert regions for a very long time. Salt is a good example. Herodotus, the Greek historian who was writing 2,500 years ago, describes the trans-Saharan salt trade and refers to an important mine that was probably the one at Taghaza, in the far north of modern-day Mali. Ibn Battuta, the Arab geographer from Tangier, described Taghaza when he visited in 1352. He thought the village was unattractive but was interested to note that its houses and mosques were constructed in blocks of salt.

Taghaza was abandoned in the 16th century, but salt for human and animal consumption is still produced at several places elsewhere in the Sahara. These include Taudenni, also in Mali, and Bilma in Niger. Salt mined from the Danakil depression in northern Ethiopia also has a lengthy history. Bars of Danakil rock salt, known as *amole* in Amharic, were used as currency in Ethiopia for 1,000 years and more. The Italians are said to have been very disappointed to find bank vaults stacked full of *amole* bars when they invaded Ethiopia in the 1930s. They were expecting to find gold. Blocks of salt are still mined by hand in the Danakil and carried on camel trains up to the Ethiopian Highlands, although the salt is used today for culinary purposes.

Another form of salt – sodium nitrate – has been exported from the hyper-arid Pacific coast of South America in more recent

20. Hacking salt from the surface of a playa in the Danakil desert, Ethiopia. Salt from the Danakil, which is thought to be 5 kilometres thick in some places, has been transported to the Ethiopian Highlands in camel caravans for many hundreds of years

times. In the late 19th century, Chilean nitrate, or 'Chile saltpetre', was used in Europe and North America as a fertilizer and in the manufacture of gunpowder. The Atacama saltpetre mines were of such geopolitical significance, and so profitable, that three countries – Chile, Peru, and Bolivia – fought over the richest of the deposits in what is called the 'Fertilizer War'. The importance of these nitrates only waned after a method for making nitrates synthetically was invented in the early 20th century.

Saltpetre was the second important form of fertilizer to come out of the Atacama. For much of the 1800s, bird droppings amassed over hundreds of years along the coastline were exported for the same purpose. This was 'guano' (derived from the Quechua word for dung), which had been used by native populations of pre-Spanish Latin America for centuries to increase crop yields. The arid climate ensured that the guano deposits, which were up to 40 metres deep on some islands off the coast of Peru, were not washed away and the dry atmosphere prevented the nitrate in the droppings from evaporating, so maintaining its effectiveness.

Gums and resins that ooze from trees in desert areas, particularly in the Middle East and the Horn of Africa, have been used by people for various purposes for a long time. The use of gum arabic, from acacia trees, can be traced back 4,000 years in Egypt, where it was used in foods, adhesives, for colouring and paint. Gum arabic has been an important cash crop for farmers in the Sahelian latitudes of Sudan ever since, and is now used all over the world.

The ancient Egyptians also sent expeditions to the 'Land of Punt', today part of Somalia, to bring back other gum resins from trees such as frankincense and myrrh. These products were transported to Europe during the times of the ancient Greeks and Romans by camel caravans along the so-called 'Incense Route' through the deserts of Arabia. They were highly prized for religious ceremonies, as well as being used to fumigate clothes, and in medicine, cosmetics, and cooking.

This trading network included the remote island of Socotra, where frankincense and myrrh were also harvested, but Socotra was probably best known in the ancient world as the sole source of cinnabar, or dragon's blood. This deep red liquid that oozes from the injured bark of the Dracaenas tree was used as a pigment in paint, to treat burns, to stain the wood for Italian violins, and even to fasten loose teeth. Roman soldiers and gladiators appreciated its disinfectant and healing properties, using it as an ointment to treat wounds.

Some of these traditional desert exports are still produced in commercial quantities, but the desert commodities used most widely in the modern world are minerals and reserves of fossil energy. Parts of the Namib desert coast are famous for their diamonds, and significant iron ore reserves are mined in the

Sahara in Mauritania and southwestern Algeria. Some of the world's largest deposits of copper are mined at Sar Cheshmeh in the arid Zagros mountains in southern Iran and at several locations in the Altiplano-Puna of South America. Deserts also provide a significant share of the world's phosphates, bauxite, and uranium.

Serious commercial exploitation of oil and natural gas in drylands dates only from the mid-19th century. Deserts in the USA and Mexico were among the first to be explored, but the world's largest reserves were discovered beneath the deserts of the Middle East and North Africa in the early 1900s. Major oilfields were found in the Arabian Gulf region following an initial strike on the fringe of the Khuzestan desert in Iran, near its border with Iraq. The Middle East is today the most important oil-producing region in the world, with Saudi Arabia, Iran, Iraq, Kuwait, and

21. Abundant sunshine is one of the most obvious desert resources, and several methods to harness solar power have been developed. This facility in the Mojave desert, southwestern USA, uses sun-tracking mirrors to heat fluid in a pipe, creating steam to drive an electrical generator

the United Arab Emirates among the largest producers. Major reserves of natural gas are also located beneath the deserts of Iran, Qatar, and Saudi Arabia particularly.

These fossil energy reserves are by definition non-renewable resources, and although the projected lifetime of the world's oil and gas fields, beneath deserts and elsewhere, depends in part on the rate at which they are used, no one doubts the fact that they will eventually run out. In the very long term, probably beyond the current century, solar power in its various forms is expected to meet most of our global energy needs. If this is indeed the case, the high levels of sunshine typical of so many deserts will again give them a central position on the world's energy scene. Technologies for harnessing solar power have advanced rapidly in recent years.

Desert ideas

Natural resources produced in desert areas have undoubtedly had significant impacts on societies far beyond the desert fringe, but the influence of ideas and lifestyles developed in deserts has been even greater. Indeed, it is no exaggeration to say that several fundamental aspects of human culture have arisen from desert beginnings. These include the domestication of plants and animals, the creation of the city, and the advent of at least three major world religions.

Some key developments in the ways human societies feed themselves have close associations with desert areas. The beginnings of farming and pastoralism were intimately linked to the process of domestication, the deliberate selection and breeding of wild plants and animals that eventually leads to their genetic change in ways that make them more suited to the needs of people. By cultivating plants, for example, farmers chose particular seeds that appeared to offer the best results and by planting more of them increased their chances of survival. Hence, the dominance of particular characteristics was deliberately increased.

The origins of domestication have been traced to diverse parts of the world using archaeological evidence which clearly shows that several important wild species from deserts and semi-deserts were among the first to undergo this process. Some of the earliest food crops to be cultivated were wheat and barley, two desert annuals, in the Fertile Crescent of the Near East around 7,000 to 9,000 years ago. Natural adaptations of these species for life in drylands made them particularly suitable for agriculture. They thrive on ephemeral supplies of water and respond by growing rapidly and producing an abundance of seeds, constituting the grain we eat.

Dryland species of livestock are similarly well suited to domestic use, being hardy and able to withstand long periods without drinking, but also capable of converting plant material into animal protein quickly and efficiently when pastures are lush. Another factor in their favour in many cases is their inclination to move naturally in herds, migrating in search of new pastures, making them amenable to shepherding. The Fertile Crescent was also where cattle, sheep, and goats were first domesticated, about 9,000 years ago.

In the Americas, llamas and alpacas were domesticated on the Andean Puna about 6,000 years ago. At around the same time, in the same place, people began to cultivate two local desert annuals: quinoa, a grain that is particularly rich in protein, and the potato.

This early transition in society, from hunting and gathering to herding and farming, probably took place for a combination of reasons. Social, economic, technological, and environmental factors were probably all involved, and this phase, the so-called 'Neolithic Revolution', was also associated with the emergence of urban civilizations on some of the world's great desert rivers (the Nile, Tigris/Euphrates, and Indus) between 3,500 and 5,500 years ago. Ancient Mesopotamia, on the Tigris/Euphrates, is generally considered to be where the first complex, urban, state-level societies emerged, in the 6th millennium before present. Urban areas developed as a result of a lengthy transition from village communities through large integrated estates to cities.

One theory that links some of the factors involved in the emergence of cities suggests that the central organization required to manage irrigation in desert areas also allowed complex societies to evolve as large numbers of people congregated to live in the same place. This tendency led eventually to the creation of the city and what are popularly thought of as the first civilizations. Another facet of this argument is the idea that the increased agricultural productivity in such areas resulted in food surpluses capable of supporting non-producing members of society, who were then free to become engaged in other activities.

Great urban economies based on irrigation certainly continued for several thousand years in the Nile Valley and in Mesopotamia, even though the origins of agriculture and urbanization remain matters for debate. The 6th and early 5th millennia before present, when the first of these highly organized and urbanized societies emerged, was also a period of great climatic and environmental change in most parts of the world. A tendency towards greater aridity has been detected in many of the regions that today contain deserts. There may be good reasons to link these environmental and social changes in Egypt, Mesopotamia, and on the Indus. This theory suggests that people responded to the drying climate by concentrating in areas with reliable supplies of water: in other words, along rivers. The suggestion is, therefore, that increased social complexity and the rise of the city was largely driven by environmental deterioration.

The process of domestication and the advent of the city are developments in human endeavour that have had the widest effects on human society. Other major contributions to global culture are the spread of three of the world's major religions, which all have their origins in the deserts of the Middle East. Judaism, Christianity, and Islam developed from desert visionaries whose profound religious experiences have each formed the basis of a faith. These three religions all recognize just one god, they are 'monotheistic', itself a concept that can be traced back to Akhenaten, an ancient ruler of the Egyptians in the Sahara.

Descendants of Abraham, patriarch of the Jewish people, left Egypt to spend 40 years in the desert until God revealed the Ten Commandments and the Torah (Judaism's religious texts) to Moses on Mount Sinai, on the Sinai Peninsula, today part of Egypt. Moses then led the Jews to the promised land of Israel.

The holy scriptures of Judaism were later referred to as the Old Testament by Christians, whose religion was founded by Jesus of Nazareth, a city located today in the state of Israel. Jesus preached his messages in the region before his execution by a prefect of the Roman Empire, but within a few hundred years Christianity had become the official religion of Rome. At around the same time, hermits had begun to form desert communities in Egypt, seeking escape from the more materialistic world of the towns and cities. One of these men, the Christian Saint Anthony, is well known for being one of the first ascetics to attempt a life in the desert completely cut off from civilization. He and his fellow 'desert fathers' had a great influence on the development of Western monasticism.

The prophet of Islam, the third major desert religion, was Muhammad, a man born in Mecca in today's Saudi Arabia. Muhammad prayed in the desert nearby where he had a vision from God on Mount Hira. Two decades after Muhammad's death, his statements and those of his immediate followers had been gathered together in the sacred text of Islam, the Koran.

Followers of Islam, like believers in Christianity and Judaism, can be found all over the contemporary world. The influence of these three 'religions of the book', each based on the profound religious experiences of desert cultures, now extends far beyond their areas of origin in sparsely populated regions of the Middle East.

Epilogue

Deserts are remarkable places. Identified by a scarcity of water and extremes of temperature, they can be harsh and hostile. But deserts are also places of great beauty and on occasion they literally teem with life. This book has been designed to demonstrate the great variety within deserts, their climates, landscapes, wildlife, and human use. Each is unique in some way, whether through fantastic life forms, extraordinary scenery, or ingenious human adaptations.

Deserts are one of the most extensive habitats on Earth, populated by a range of creatures that stretches from tiny algae to great mammals. These animals, along with dryland plants, have by necessity become adapted to the great variety of precipitation patterns, temperature regimes, topographies, and environmental histories that make up the dryland realm. Some deserts also appear to be particularly sensitive to changes in climate, a characteristic that should interest us today because a billion people live in drylands, but also because any alterations to the ways deserts function may serve as an early warning to changes that will occur in other environments.

Deserts have been important to people in innumerable ways. In the widest sense, they form an integral part of the planet we inhabit. This is evident from the many connections between deserts and the rest of the world, both human and physical. Deserts are also significant for many more immediate reasons. People have lived in and around them for a very long time. Through the ages, desert resources have been prized far beyond the dryland sphere, with examples ranging from frankincense to crude oil.

The history of human societies also shows us that deserts have spawned a surprisingly large number of ideas and innovations that have become central to the way human societies work everywhere. We have learned from our experiences in deserts, and we should continue to do so. Some modern communities, particularly in cities, are in danger of forgetting some of these lessons. We can divorce ourselves from the vagaries of the natural environment with technology, but doing so also entails an element of risk. Cutting ourselves off from Nature altogether is not possible. It might be wiser to admire rather than reject the resilience and adaptability of some more traditional desert lifestyles.

Perhaps above all, many of the world's great desert areas remain as havens of peace and solitude in a world bursting with humanity. Beneath blazing sun or suffocating dust, huge, seemingly virgin landscapes stretch as far as the eye can see and much further. They are exceptional places, evocative of a time before people, when Nature was raw and the world was made up only of wilderness.

Further reading

D. Chatty (ed.), *Nomadic Societies in the Middle East and North Africa: Entering the 21st Century* (Leiden: Brill, 2006).

W. R. J. Dean and S. Milton (eds.), *The Karoo: Ecological Patterns and Processes* (Cambridge: Cambridge University Press, 1999).

A. de Saint Exupery, *Wind, Sand and Stars* (New York: Mariner Books, 1939).

E. Ezcurra (ed.), *Global Deserts Outlook* (Nairobi: United Nations Environment Programme, 2006).

A. Goudie, *Great Warm Deserts of the World* (Oxford: Oxford University Press, 2002).

C. F. Hutchinson and S. M. Herrmann, *The Future of Arid Lands Revisited: A Review of 50 Years of Drylands Research* (Paris: UNESCO and Dordrecht: Springer, 2008).

J. Laity, *Deserts and Desert Environments* (Chichester: Wiley-Blackwell, 2008).

P. Latz, *Bushfires and Bushtucker: Aboriginal Plant Use in Central Australia* (Alice Springs: IAD Press, 1995).

M. A. Mares (ed.), *Encyclopedia of Deserts* (Norman: University of Oklahoma Press, 1999).

C. Moss, *Patagonia: A Cultural History* (Oxford: Signal Books, 2008).

A. J. Parsons and A. D. Abrahams (eds.), *Geomorphology of Desert Environments*, 2nd edn. (Berlin: Springer, 2009).

B. A. Portnov and P. A. Hare (eds.), *Desert Regions: Population, Migration and Environment* (Heidelberg: Springer-Verlag, 1999).

J. F. Reynolds and D. M. Stafford Smith (eds.), *Global Desertification: Do Humans Cause Deserts?* (Dahlem Workshop Report 88; Berlin: Dahlem University Press, 2002).

G. B. Schaller, *Wildlife of the Tibetan Steppe* (Chicago: University of Chicago Press, 2000).

M. Shostak, *Nisa: The Life and Words of a !Kung Woman* (New York: Vintage Books, 1981).

M. Smith and P. Hesse (eds.), *23°S Archaeology and Environmental History of the Southern Deserts* (Canberra: National Museum of Australia Press, 2005).

W. Thesiger, *Arabian Sands* (London: Penguin, 1959).

D. S. G. Thomas and N. Middleton, *Desertification: Exploding the Myth* (Chichester: Wiley, 1994).

P. Veth, M. Smith, and P. Hiscock (eds.), *Desert Peoples: Archaeological Perspectives* (Chichester: Wiley-Blackwell, 2005).

D. Ward, *The Biology of Deserts* (Oxford: Oxford University Press, 2009).

D. Worster, *Dust Bowl: The Southern Plains in the 1930s* (New York: Oxford University Press, 1979).

Deserts

Index

Index

Deserts

Index

Expand your collection of
VERY SHORT INTRODUCTIONS

ONLINE CATALOGUE
A Very Short Introduction

Our online catalogue is designed to make it easy to find your ideal Very Short Introduction. View the entire collection by subject area, watch author videos, read sample chapters, and download reading guides.

http://fds.oup.com/www.oup.co.uk/general/vsi/index.html

SOCIAL MEDIA
Very Short Introduction

Join our community

www.oup.com/vsi

- Join us online at the official Very Short Introductions **Facebook** page.
- Access the thoughts and musings of our authors with our online **blog**.
- Sign up for our monthly **e-newsletter** to receive information on all new titles publishing that month.
- Browse the full range of Very Short Introductions online.
- Read **extracts** from the Introductions for free.
- Visit our library of **Reading Guides**. These guides, written by our expert authors will help you to question again, why you think what you think.
- If you are a teacher or lecturer you can order inspection copies quickly and simply via our website.